ELECTRONIC MEDIATIONS
KATHERINE HAYLES, MARK POSTER, AND
SAMUEL WEBER, SERIES EDITORS

Cyberculture

Pierre Lévy

Translated by Robert Bononno

Electronic Mediations, Volume 4

University of Minnesota Press
Minneapolis • London

The University of Minnesota Press gratefully acknowledges financial assistance provided for the translation of this book by the French Ministry of Culture and by the McKnight Foundation.

Originally published in French under the title *Cyberculture*, copyright 1997 Éditions Odile Jacob/Éditions du Conseil de l'Europe.

Published by the University of Minnesota Press
111 Third Avenue South, Suite 290
Minneapolis, MN 55401-2520
http://www.upress.umn.edu

Printed in the United States of America on acid-free paper

Library of Congress Cataloging-in-Publication Data

Lévy, Pierre, 1956–
　[Cyberculture. English]
　Cyberculture / Pierre Lévy ; Translated by Robert Bononno.
　　p. cm.—(Electronic mediations ; v. 4)
Includes bibliographical references and index.
　ISBN 0-8166-3609-5 (hardback : alk. paper)—ISBN 0-8166-3610-9 (pbk. : alk. paper)
　1. Information technology—Social aspects. 2. Telecommunication—Social aspects.
3. Internet—Social aspects. 4. Cyberspace—Social aspects. 5. Culture.
6. Computers and civilization.
I. Title. II. Series
　HM851 .C9313 2001
　303.48′33—dc21
　　　　　　　　　　　　　　　2001001664

10 09 08 07 06 05 04 03 02 01　10 9 8 7 6 5 4 3 2 1

To my parents, Lilia and Henri

Contents

Part III. Problems

Deluges

This book is an attempt to understand and explain cyberculture. I am generally—and correctly—considered to be an optimist. But in spite of my optimism, I do not believe that the Internet will magically resolve all of the planet's social and cultural problems. I do, however, want to acknowledge the following: First, that the growth of cyberspace is the result of an international movement of young people eager to experiment collectively with forms of communication other than those provided by traditional media. Second, that a new communications space is now accessible, and it is up to us to exploit its most positive potential on an economic, political, cultural, and human level.

Those who denounce cyberculture today strangely resemble those who criticized rock music during the fifties and sixties. Rock started out as an Anglo-American phenomenon and has become an industry. Nonetheless, it was able to capture the hopes of young people around the world and provided enjoyment to those of us who listened to or played rock. Sixties pop was the conscience of one or two generations and helped bring the war in Vietnam to a close. Obviously, neither rock nor pop has solved global poverty or hunger. But is this a reason to be "against" them?

During one of the many frequent roundtables organized to assess the "impact" of new communications networks, I had the opportunity to listen to a former filmmaker—now a European bureaucrat—denounce the "barbarism" inherent in video games, virtual worlds, and electronic forums. I remarked that this was a strange comment from a representative of the cinema. During its formative years, film was also condemned as a mechanical tool for "dumbing down" the masses by nearly every well-meaning intellectual and cultural spokesperson. Yet today film is recognized as a legitimate art and a legitimate form of cultural expression. Unfortunately, it appears as if we haven't learned much from our mistakes. The same phenomena that were at work in cinema have been reproduced in social and artistic forms of expression based on contemporary technologies, which are denounced as "foreign" (American), inhuman, inane, escapist, et cetera.

I don't wish to imply that everything that takes place on digital networks is "good." That would be as absurd as claiming that all films are works of art. I'm simply asking that we remain open to innovation, that we make an attempt to understand it. The real issue is obviously not whether we are for or against technology but whether we recognize the qualitative changes in the ecology of signs, the unfamiliar environment that results from the extension of new communications networks throughout social and cultural life. Only then will we be able to develop these new technologies within a humanist perspective.

Are dreamers the only ones who speak of humanism? The issues are clear enough. Newspapers and television have clarified them for us: cyberspace has entered the realm of commerce. "Businessmen Mount Assault on the Internet," to quote a headline from *Le Monde diplomatique*. Now it's simply a question of money. Activism and utopianism are things of the past. When we attempt to explain the development of new non-hierarchic, interactive, and cooperative forms of communication, we're reminded of the otherworldly profits made by Bill Gates, CEO of Microsoft. On-line services will be billed, reserved for the wealthiest. The growth of cyberspace will do nothing more than widen the gap between the haves and have-nots, between the Northern Hemisphere and the poor regions of the earth where the majority of inhabitants don't even have a telephone. Any attempt to understand cyberculture automatically makes you an ally of IBM, international market capitalism, and the U.S. government, an apostle of crude neoliberalism and enemy of the poor, a proponent of globalization beneath the mask of humanism.

I would like, therefore, to offer a few commonsense arguments about technology. That cinema and music are industries and a source of revenue does not prevent us from enjoying them or speaking from a cultural or aesthetic perspective. The telephone has proved to be a source of tremendous wealth for telecommunications companies. This in no way mitigates the fact that telephone networks enable planetary and interactive communication. Nor is the fact that only a quarter of humanity has access to telephones an argument "against" the telephone. There is no reason, therefore, why the economic exploitation of the Internet or the failure to provide global access should in themselves serve as a condemnation of cyberculture or prevent us from approaching it other than with a critical eye. Admittedly there are an increasing number of paid services on the Internet. And it would appear that this development is going to continue and probably accelerate in the next few years. Nonetheless we mustn't lose sight of the fact that free services have been growing even more rapidly. These services are provided by universities, government and nonprofit organizations, individuals, interest groups, even corporations. There is no reason to separate commerce from the libertarian and communitarian dynamic that presided at the birth of the Internet. The two are complementary.

The matter of exclusion is obviously crucial and will be a subject I'll return to in the final chapter. I would simply like to point out here that it should not prevent us from anticipating the cultural implications of cyberculture in all its dimensions. Moreover, it's not the poor who are "against the Internet," but those whose power, privilege (especially cultural), and monopoly are threatened by the emergence of new configurations of the communications infrastructure.

During an interview that took place in the fifties, Albert Einstein remarked that three bombs with global repercussions had exploded during the twentieth century: the demographic bomb, the atomic bomb, and the telecommunications bomb. What Einstein referred to as the "telecommunications bomb," my friend Roy Ascott (one of the pioneers and principal theoreticians of network art) called the second, or information, deluge. Telecommunications engendered this second deluge because of the exponential, explosive, and chaotic nature of its growth. The amount of data available is multiplying at an accelerating rate. The density of the links among information sources is increasing at a dizzying pace within data banks, hypertexts, and networks. Nonhierarchical contacts among individuals are proliferating anarchically. The result is a chaotic overflow of information, a flood of data swept along by the tumultuous, rolling waters of communication, the deafening cacophony and repetition of the media, a war of images, propaganda and counter-propaganda, intellectual confusion.

The demographic bomb is also a kind of deluge, an unimaginable wave of people. There were slightly more than 1.5 billion people on earth in 1900; there were nearly 6 billion by 2000. Humanity is overflowing the earth. Such rapid and widespread global growth is without historical precedent.

In the face of this unstoppable human flood, two solutions are possible. The first is war, extermination in an atomic deluge, regardless of the form it takes and the contempt for human life it implies. In this case, human life loses its value. The human is reduced to the level of a farm animal or an ant: hungry, terrorized, exploited, deported, massacred.

The second is the exaltation of the individual, the human considered as the principal source of value, a marvelous and priceless resource. To enhance this value, we tirelessly endeavor to weave relationships between generations, sexes, nations, and cultures, in spite of the difficulties, in spite of the conflicts. This second solution, symbolized by telecommunications, implies the recognition of the *other*, mutual acceptance, assistance, cooperation, association, and negotiation, beyond our divergent viewpoints and interests. From one end of the world to the other, telecommunications extends the possibilities of amicable contact, contractual transactions, the transmission of knowledge and exchange of understanding, the pacific discovery of difference.

The delicate web of humanity, a single, immense fabric, open and interactive, results in a situation without precedent, one that is not only full of hope because it is a positive response to demographic growth but the creator of new kinds of problems. I would like to address some of those problems in this book, particularly those that relate to culture: art, education, and political life as conditioned by generalized inter-active communication. At the dawn of the information deluge, some thoughts on the biblical flood might help clarify our contemporary situation. Where is Noah? What should he put in the ark?

In the midst of the surrounding chaos, Noah constructed a small, well-ordered whole. He protected a sample of the data that surged around him. As his world fell apart, he was seized with the desire to transmit. And in spite of the general madness, he gathered his samples with an eye on the future.

"And the Lord shut him in" (Genesis 7:16). The ark was sealed, symbolizing a reconstituted totality. With the universe in turmoil, the organized microcosm reflects the order of a macrocosm yet to come. But there is no mistaking multiplicity. The flood of information will never subside. The ark will never come to rest on Mount Ararat. The second deluge will never end. There is no solid foundation beneath the ocean of information. We must accept it as our new condition, one in which our children will learn to swim, float, and navigate.

When Noah, that is, each of us, looked through the porthole of his ark, he saw other arks in the far distance, floating on the howling ocean of digital communication. Each of those arks contained something different. Each wanted to save diversity. Each wanted to transmit. Those arks will wander indefinitely on the surface of the waters.

One of the principal hypotheses of this book is that cyberculture expresses the rise of a new universal, different from the cultural forms that preceded it because it is constructed from the indeterminateness of some global meaning. It is important, however, that we try to situate it within the perspective of the previous mutations of communication.

In oral societies, discursive messages are always received in the context in which they are produced. But with the arrival of writing, texts are detached from the context in which they are created. We can read a written message five centuries after it was written or five thousand kilometers away, which often introduces problems of reception and interpretation. To overcome these difficulties, some types of messages are specially designed to retain the same meaning regardless of the context (place or time) in which they are received: these are "universal" messages (science, scripture, human rights, etc.). This universality, acquired through static writing, can only be constructed at the cost of a certain closure or fixity of meaning. The universal is totalizing. My hypothesis is that cyberculture reinstates the copresence of messages and their context, which had been current in oral societies, but on a different scale and on a different plane. The new universality no longer depends on self-sufficient texts, on the fixity and independence of signification. It is constructed and extended by interconnecting messages with one another, by their continuous ramification through virtual communities, which instills in them varied meanings that are continuously renewed.

The ark of the first deluge was unique, watertight, sealed, totalizing. The arks of the second deluge dance in concert. They exchange signals, impregnate one another. They shelter small totalities, but without any claim to universality. The deluge alone is universal. But it can't be totalized. Noah has grown modest.

"And every living substance was destroyed ... and Noah only remained alive, and they that were with him in the ark" (Genesis 7:23). Noah's rescue operation appears to be complementary, almost an accessory, to extermination. The would-be universal totality drowns everything it does not contain. That is how civilizations are founded, how an imperial universal arises. In China, the Yellow Emperor had nearly all the texts that existed before his reign destroyed. What Caesar, what barbarian conqueror, allowed the library of Alexandria to burn so that the disorder of Hellenism would be wiped away once and for all? The Spanish Inquisition and its autos-da-fé allowed the Koran, the Talmud, and untold thousands of pages of inspired meditation to go up in smoke. Hitler's bonfires destroyed books throughout the cities of Europe, consuming intelligence and culture in the conflagration. The first of all attempts at destruction may have been the one that took place in Mesopotamia, the oldest empire and birthplace of writing, whose legend of the flood predates that in the Bible. Wasn't it Sargon of Agade, king of the Four Regions, the first emperor in history, who caused thousands of clay tablets, engraved with timeless legends, precepts of wisdom, manuals of medicine and magic, produced by generations of scribes, to be thrown into the Euphrates? Briefly visible beneath the river's surface, the signs eventually disappeared. Tossed about by the turbulent waters, polished smooth by the current, the tablets slowly began to soften, once again became smooth pieces of clay that would soon dissolve in the mud and return to the silt from which they originated. Voices were silenced for time immemorial. There would be no echo, no response.

But the new deluge cannot erase the marks of the intellect. It sweeps everything along in its path. Fluid, virtual, simultaneously gathered and dispersed, it is impossible to burn this library of Babel. The innumerable voices that resonate through cyberspace will continue to call and respond to one another. In this deluge, the floodwaters will never wash away the signs that have been engraved.

Yes, technology has produced nuclear destruction along with interactive networks. But the telephone and the Internet can "only" be used for communication. For the first time in this century of iron and idiocy, the atom and telecommunications have succeeded in *unifying the human race*: death on a specieswide scale in the case of the atomic bomb, planetary dialogue in the case of telecommunications.

Technology is responsible for neither our salvation nor our destruction. Always ambivalent, technologies project our emotions, intentions, and projects into the material world. The instruments we have built do provide us with power, but since we are collectively responsible, the decision on how to use them is in our hands.

This book, originally commissioned by the Council of Europe, focuses on the cultural implications of the development of digital technologies of information and communication. Economic and industrial factors, the problems of employment, and legal questions have been excluded from the scope of this study. Instead, I want to focus on our general attitude in the face of new technologies, the current virtualization of information and communication, and the resulting global mutation of civilization. I have paid particular attention to new forms of art, the transformation of our relationship to knowledge, the implications for education and training, urban life and democracy, the maintenance of the diversity of languages and cultures, the problems of exclusion and inequality.

Because I frequently use the terms "cyberspace" and "cyberculture," I thought it would be useful to provide a brief definition. Cyberspace (also known as the "network") is the new medium of communications that arose through the global interconnection of computers. The term refers not only to the material infrastructure of digital communications but to the oceanic universe of information it holds, as well as the human beings who navigate and nourish that infrastructure. Cyberculture is the set of technologies (material and intellectual), practices, attitudes, modes of thought, and values that developed along with the growth of cyberspace.

In Part I, where I introduce the question of the social and cultural impact of new technology, I provide a summary description of the major technological concepts that express and support cyberculture. While reading them, the reader should bear in mind that these technologies are creating new conditions and provide unimaginable opportunities for the development of the individual and society. However, they do not automatically determine whether humanity's future will be hidden by darkness or bathed in light. I have attempted to provide the clearest definitions possible, for although the topic is now well known to the general public, it is often understood piecemeal, without the precision and clarity that are essential to an understanding of the major forces affecting society. I have tried to provide nonspecialists with access to the concepts of digitization, hypertext and hypermedia, computer simulations, virtual reality, the major features of interactive networks, and especially the Internet.

Part II covers the cultural implications of the development of cyberspace. It paints a portrait of cyberculture: the new form of universality it invents, the social movement in which it arose, its artistic and musical genres, the upheaval it has created in our relation to knowledge, the necessary educational reform it has caused, its contribution to urbanism and our notions of the city, the questions it raises for political philosophy.

Part III explores the negative side of cyberspace through the antagonisms and criticisms it has managed to provoke. Here I discuss the conflicts of interest and power struggles that take place in and around cyberspace, the often virulent denunciations of the virtual, the question of exclusion, and the maintenance of cultural diversity in the face of political, economic, and media imperialism.

Cyberculture was originally published in French in 1997. Since then the Internet and questions about its use and future development have continued to evolve. Nonetheless, I feel that the fundamental philosophical issues I examine here are as valid now as they were then.

P. L.

June 2000

PART I
DEFINITIONS

CHAPTER 1

The Impact of Technology

The Metaphor of Impact

In conference announcements, abstracts of official studies, and articles by the mass media concerning the development of multimedia, the question of the "impact" of new information technology on society or culture is often raised. Technology is compared to a projectile (a stone, shell, or missile), and culture or society to a living target. This ballistic metaphor can be criticized on more than one account. It is not so much a question of evaluating the stylistic relevance of a rhetorical figure as it is of updating the way we read phenomena—inadequate as far as I'm concerned—revealed by the impact metaphor.[1]

Does technology come from another planet, a cold, unemotional world of machines, foreign to any signification or human value, as some critical traditions have suggested?[2] On the contrary, it seems to me that not only is technology imagined, fabricated, and reinterpreted for use by humankind, but it is this intensive use of tools that constitutes our humanity as such (together with language and complex social institutions). The man who speaks, buries his dead, and chips away at a block of silex is one and the same. From past to present, the fire of Prometheus has cooked our food, baked clay, melted metals, powered steam engines, coursed through high-voltage cables, burned in nuclear reactors, and exploded in our bombs and weapons of destruction. Through the architecture that shelters, gathers, and inscribes it on Earth; through the wheel and navigation, which have expanded its horizons; through writing, the telephone, and cinema, which infiltrate it with signs; through text and textile, which, as they weave together a variety of materials, colors, and meanings, unfurl its undulating surfaces, the luxurious folds of its intrigues, fabrics, and veils—the human world is technological to its core.

Is technology an autonomous factor, separate from society and culture, which are no more than passive entities pierced by some external agent? It is my contention that technology is a way of analyzing global sociotechnological systems, a point of view that emphasizes the material and artificial components of human phenomena, and not a real entity, which exists independently, has distinct effects, and acts on its own. Human activities comprise indissoluble interactions among

- living and thinking beings
- natural and artificial material entities
- ideas and representations

It is impossible to separate the human from its material environment, or from the signs and images through which humanity gives meaning to life and the world. Similarly, we cannot separate the material world—even less so its artificial component—from the ideas through which technological objects are conceived and used, or from the humans who invent, produce, and use them. Moreover, images, words, and linguistic constructions reside in the human mind, providing humankind and its institutions with means and reasons for living, and are transported in turn by organized, tool-bearing groups, just as they are by communications circuits and artificial memories.[3]

Even if we assume the existence of three entities, technology, culture, and society, rather than emphasizing the impact of technology, we could also treat technology as the product of a society and a culture. But the distinction between culture (the dynamic of representation), society (people, their interrelations, exchanges, relationships of force), and technology (useful artifacts) can only be conceptual. There is no corresponding agent, no truly independent "cause." We have an intellectual bias for agents because there are real groups associated with these linguistic elements (ministries, scientific disciplines, university departments, research laboratories) or because certain forces want us to believe that a given problem is "purely technological" or "purely cultural" or "purely economic." There are therefore no genuine relationships between "a" technology (part of the cause) and "a" culture (which would undergo its effects), but among a multitude of human agents who variously invent, produce, use, and interpret *technologies*.[4]

Technology or Technologies

Technologies embody projects, imaginary schemes, highly varied social and cultural implications. Their presence and use in a given place, at a given time, crystallizes the changing relations of force among human beings. The steam engine enslaved workers in the textile factories of the nineteenth century, whereas the personal computer increased the individual's ability to act and communicate during the twentieth. We cannot, in effect, speak of the sociocultural effects or the meaning of technology in general, as Heidegger's[5] followers and indeed the entire tradition based on the Frankfurt school[6] have tried to do. Is it legitimate, for example, to compare nuclear science and electronics directly? One is limited to centralized organizations controlled by specialists, imposes strict standards of safety and security, and commits us to making long-term decisions. Electronics, which is much more versatile, is as appropriate to hierarchical organizations as it is to the wide-scale distribution of power and follows much shorter techno-economic cycles.[7]

Behind the technologies, ideas, social projects, utopias, economic interests, and strategies of power—the entire range of humankind's activities in society—can be seen acting and reacting. Therefore any assignment of a univocal meaning to technology is questionable. The ambivalence and multiplicity of the significations and projects enveloping technology are especially evident in the case of digital technology. The development of cybertechnology is encouraged by governments, who are in search of power in general and military supremacy in particular. It is also a major factor in global competition among the large electronics and software companies, and the major geopolitical groups. Yet it also responds to the needs of the designers and users, who are looking to increase individual autonomy and expand their cognitive faculties. It embodies the ideal of scientists, artists, managers, and network activists who want to improve collaboration among individuals, who are exploring and bringing to life different forms of distributed and collective intelligence. Although these heterogeneous projects are occasionally in conflict, they more frequently nourish and mutually reinforce one another.

The difficulty of analyzing the social and cultural implications of information technology or multimedia is multiplied by the complete lack of stability in the field. Aside from the logic principles used as the basis for computational science, what common ground is there between the behemoths of the fifties, which were used for scientific calculations and statistics, filled entire rooms, were very expensive, and lacked monitors and keyboards, and the personal computers of the eighties, which required no scientific or technical training, and could be used by nearly anyone to write, draw, compose music, or plan a budget? They are both computers, but the cognitive, cultural, economic, and social implications are obviously quite different. Yet digital technology is still at the beginning of its trajectory. The global interconnection of computers (the extension of cyberspace) continues unabated. The future standards for multimode communications are now being fought over. Tactile, auditory, capable of interactive three-dimensional visualization, the new interfaces with the digital universe are becoming increasingly commonplace. To help us navigate within the flow of information, today's laboratories compete with one another in designing printed circuit boards for dynamic data transmission and developing intelligent software agents, or *knowbots*. These phenomena are transforming the cultural and social significations of cybertechnology.

Given the scope and rate of past transformations, it is still impossible to predict the changes that will affect the digital universe after 2000. Whenever computer memory and bandwidth increase, whenever we invent new interfaces to the human body and its cognitive system (such as virtual reality), whenever we translate the content of old media into their cyberspace equivalents (telephone, television, newspapers, books), whenever digital technology enables formerly separate physical, biological, psychic, economic, and industrial processes to communicate with one another, their social and cultural implications must be reevaluated.

Does Technology Determine or Condition?

Does technology determine society and culture? If we accept this fiction, we find that the relation is far more complex than one of determination. The emergence of cyberspace accompanies, translates, and promotes the general evolution of civilization. A technology is produced within a culture, and a society is conditioned by its technologies. *Conditioned*, not determined. The difference is critical. The invention of the stirrup led to the development of new forms of heavy cavalry, which served as the basis for the rise of chivalry and the political and social structures of feudal society. Yet the stirrup, as a material device, is not the "cause" of European feudalism. There is no identifiable cause for a social or cultural state of affairs, but an infinitely complex and partially indeterminate set of interacting processes, which strengthen or inhibit one another. Nevertheless, without the stirrup, it's hard to see how a knight in armor could have remained upright on his horse and charged with his lance. The stirrup did indeed condition chivalry and, indirectly, all feudalism, yet it did not determine them.

To say that technology conditions is to imply that it provides access to certain possibilities, that certain cultural or social options couldn't seriously be contemplated without its presence. Yet several possibilities remain open, and many are left untouched. The same technology can be integrated into vastly different cultural circumstances. For example, although large-scale crop irrigation may have favored "Oriental despotism" in Mesopotamia, Egypt, and China, all three civilizations are very different, and irrigation was often adapted to cooperative sociopolitical forms (such as the medieval Maghreb). Printing, which was banned in China but as an industrial activity escaped political control in Europe, did not have the same consequences in the East and the West. Gutenberg's press did not determine the crisis of the Reformation, the development of modern European science, or the rise of Enlightenment idealism and the growing power of public opinion in the eighteenth century; it only conditioned them. It remained an essential element of the global environment in which these cultural forms arose. Although an intransigent mechanistic philosopher might claim that an effect is determined by its causes and can be deduced from them, simple common sense suggests that cultural and social phenomena do not obey this pattern. The multiplicity of factors and agents inhibits the calculation of deterministic effects. Moreover, "objective" factors are never more than conditions to be interpreted by people and collectivities capable of radical invention.

A technology is neither good nor bad (depending on context, use, and point of view), or even neutral, for that matter (since it conditions or constrains, exposes or closes off, the range of possibilities). It is a question not of evaluating its "impact" but of identifying those points of irreversibility where technology forces us to commit ourselves and provides us with opportunities, of formulating the projects that will exploit the virtualities it bears within it and deciding what we will make of them.

Yet it would be illusory to believe in the total availability of technologies and their potential for supposedly free and rational individuals or collectivities. Often by the time we begin to deliberate on the possible uses of a given technology, those uses are already being imposed on us. Even before the rise of conscious awareness, the collective dynamic has already established its attractors. By the time we begin to pay attention to a technology, it is too late; the moment has passed. While we are still questioning its utility, other technologies are emerging on the nebulous frontier where ideas, things, and practices are invented. Still invisible to us, they may be on the verge of disappearance or destined for success. Within these zones of indeterminacy where our future is played out, groups of marginal designers, enthusiasts, and bold entrepreneurs struggle to inflect our becoming. No important institutional agent—government or corporation—deliberately planned, no major media was able to predict, the development of personal computing or the widespread use of interactive graphical interfaces, or bulletin boards, or the software programs that would make possible virtual communities, hypertext, and the World Wide Web, or unbreakable personal cryptography.[8] These technologies, all of which are impregnated with the marks of first use and their designers' intentions, born within the context of a visionary spirit and carried forward by the ferment of social movements and grassroots practices, entered our field of vision from a place no "decision maker" expected them to be.

The Acceleration of Technological Change and Collective Intelligence

If we focus on its signification for humanity, it appears that, as I suggested earlier, digital technology—fluid and in constant motion—lacks any stable essence. Yet paradoxically, the rate of change is itself a constant of cyberculture. It partly explains the sensation of impact, exteriority, and strangeness that grips us whenever we attempt to apprehend the contemporary movement of technology. For the individual whose method of working is suddenly modified, for a profession that is suddenly touched by a technological revolution that renders its traditional skills obsolete (typographers, bankers, pilots)—or threatens its very existence—for the social classes or regions of the world that do not participate in the excitement of the design, production, or ludic appropriation of new digital instruments, technological evolution appears as the manifestation of a threatening "other." In truth, each of us experiences this state of dispossession at one time or another. The acceleration is so strong and so general that even the most "sophisticated" individuals find themselves overtaken by change, since no one can actively participate in transforming the entire range of technological specializations or even follow them closely.

What we crudely label "new technologies" encompasses the multiform activity of human groups, a complex collective becoming that crystallizes around material objects, computer programs, and communications devices. It is the social process in all its opacity, the *activity of the other*, which returns to the individual in the form of the foreign, inhuman mask of technology. When the impact of technology is negative, we should question the organization of labor, or relations of domination, or the tangled complexity of social phenomena. Similarly, when the impact is felt to be positive, it is obviously not technology that is responsible for success, but the people who conceived, implemented, and employed its instruments. In this case, the quality of the process of appropriation (that is, ultimately, the quality of human relationships) is often much more important than the systemic particularity of our tools, to the extent that these can be separated.

The faster technology changes, therefore, the more it seems to come from somewhere outside. Moreover, the feeling of strangeness increases with the separation of activities and the opacity of social processes. It is here that the central role played by collective intelligence is felt most strongly, for it is one of the principal engines of cyberculture.[9] The synergy of skills, resources, and projects, the constitution and dynamic maintenance of shared memories, the activation of flexible and nonhierarchical modes of cooperation, the coordinated distribution of decision centers stand in sharp contrast to the hermetic separation of activities, the insularity and opacity of social organization. As the process of collective intelligence develops—which quite obviously calls into question relations of power—individuals and groups will more easily appropriate technological change, and the ability of accelerating technosocial movements to cause human destruction and exclusion will diminish. Yet cyberspace, an interactive and community-based means of communication, is one of the key instruments of collective intelligence. It is through cyberspace, for example, that professional training and educational organizations have developed cooperative learning programs based on the use of interconnected networks. Large corporations have implemented computerized mechanisms for collaboration and decentralized coordination (groupware). Researchers and students throughout the world exchange ideas, articles, images, experiences, and observations in electronic forums organized around shared interests. Computer scientists across the globe help one another solve programming problems. Specialists help novices and are helped in turn by other specialists in fields with which they are unfamiliar.

Collective Intelligence: Poison and Cure

As the medium for collective intelligence, cyberspace is one of the principal conditions for its own development. The history of cyberculture bears witness to this process of positive feedback, the self-maintaining aspect of the revolution in digital networks.[10] The phenomenon is both complex and ambivalent.

The growth of cyberspace does not automatically determine the development of collective intelligence; it only provides a friendly environment for it. Indeed, we find a wide variety of new forms coming to life within the orbit of interactive digital networks:

- cognitive isolation and overloading (the stress of communication and computer use)
- dependence (addiction to the Web or games within virtual worlds)
- domination (reinforcement of decision and control centers, quasi-monopolistic control by economic powers of important network functions)
- exploitation (certain instances of monitored telecommuting or the delocalization of activities in the Third World)
- *collective stupidity* (rumors, the conformity of the network and virtual communities, the accumulation of data devoid of information, "interactive television")

In those cases where the processes of collective intelligence have developed effectively through cyberspace, their effect has been to again accelerate the rate of technosocial change. This makes our active participation in cyberculture even more compelling. Nonparticipants will be left behind, and those who have yet to enter the positive cycle of change, including its comprehension and appropriation, will be excluded even more radically than before.

Because it is participatory, socializing, decompartmentalizing, and emancipating, the collective intelligence offered by cyberculture is one of the best remedies for the destabilizing, and often exclusionary, effects of the rate of technological change. At the same time, collective intelligence works actively to accelerate this mutation. In ancient Greece, the word *pharmakon* (which gave us the word "pharmacy") referred to both a poison and a remedy. A new *pharmakon*, the collective intelligence promoted by cyberculture acts as a *poison* for nonparticipants (and no one individual can participate completely because of its size and diversity) and a *remedy* for those who are willing to plunge into its turbulence and negotiate its currents.

CHAPTER 2

The Technical Infrastructure
of the Virtual

The Emergence of Cyberspace

The first computers (preprogrammed calculators) appeared in England and the United States in 1945. Initially they were used solely by the military to perform scientific calculations, but civil use expanded throughout the sixties. Even at this time it was clear that the performance of computer hardware would increase at a steady pace. Yet aside from a handful of visionaries, no one at the time suspected that a general movement toward the virtualization of information would so profoundly affect the fundamental data of social life. Computers were still big calculators, fragile, isolated in refrigerated rooms, into which scientists in white smocks fed punched cards and out of which unreadable lists of data were periodically disgorged. Computer technology was used for scientific calculations, government and business statistics, and large-scale administrative functions (payroll, accounting, etc.).

The turning point occurred during the 1970s. The development and commercialization of the microprocessor (a device for performing logical and arithmetic calculations, lodged on a single small electronic chip) triggered several large-scale economic and social processes. The seventies gave rise to a new phase of industrial development—robotics, flexible manufacturing cells, numerically controlled machine tools—and saw automation introduced for the first time in the service industries (banking, insurance). Since then the systematic search for productivity gains through the use of electronic devices, computers, and communications networks has gradually taken over all economic activities. This trend continues to this day.

At the same time, a new social movement was taking shape in California. Fed by the ferment of the counterculture, it embraced the new technological developments and gave rise to the personal computer. From then on, the computer would gradually begin to escape the grasp of business information services and professional programmers to become an instrument of creation (texts, images, music), organization (databases and spreadsheets), simulation (spreadsheets, decision-making tools, research software), and entertainment (games) in the hands of an ever larger proportion of the population of the developed countries.

During the eighties, the contemporary horizon of multimedia took shape. Slowly, computer science lost its technological and industrial aspect and began to merge with telecommunications, publishing, cinema, and television. Digitization first became commonplace in music production and recording, but microprocessors and computer memory soon became an integral part of the infrastructure of every field of communications. New kinds of "interactive" messages began to appear: the decade witnessed the arrival of video games, the triumph of user-friendly computing (graphical interfaces and sensorimotor interactions), and the appearance of hyperdocuments (hypertext, CD-ROM).

Toward the end of the eighties and the beginning of the nineties, a new social and cultural movement made up of young professionals in large cities and college campuses across America quickly spread throughout the world. Although there was no central controlling body, the various computer networks that had come into existence since the end of the seventies were joined together, and the number of people and computers connected to this internetwork began to grow exponentially. As with the creation of the personal computer, a spontaneous and unforeseeable cultural current imposed a new direction on technical and economic development. Digital technologies became the infrastructure of cyberspace. The new space served as the locus of communication, sociability, organization, and transaction, as well as a new market for the exchange of information and knowledge.

In this context, technology in and of itself is not very interesting. It is, however, important to outline the major trends in the evolution of contemporary technology before examining the social and cultural mutations that accompanied it. Perhaps the most important element in this scenario is the exponential increase in hardware performance (computational speed, storage capacity, bandwidth) combined with continuously dropping prices. At the same time, software has benefited from the conceptual and theoretical improvements that have accompanied the increase in hardware performance. Today software publishers are committed to constructing a work and communications environment that is increasingly transparent and user-friendly.

Projections regarding the social uses of the virtual must incorporate this continuous trend toward ever greater performance, lower cost, and increasing popularity. There is every reason to believe that these three trends will continue in the future. Still, it is impossible to predict the qualitative changes that will accompany these movements or the way in which society will incorporate or inflect them; for at this point, conflict may occur among divergent projects, in which technological, economic, and social forces are indissolubly mixed.

Processing

Information technology encompasses all the technologies that are used to digitize information (input), store it (memory), process it automatically, transport it, and put it at the disposal of an end user, whether human or machine (output). These distinctions are conceptual. The actual hardware and components always combine several functions.

Information-processing units, or processors, which reside on "chips," perform arithmetic and logic calculations on data. They carry out at high speed, and in a highly repetitive fashion, a small number of very simple operations on digitally encoded information. From tubes to transistors, from transistors to integrated circuits, from integrated circuits to microprocessors, extremely rapid advances in information processing have benefited from improvements in circuit architecture, progress in electronics and physics, and applied research on materials. Each year processors become smaller, more powerful, more reliable, and less expensive.[1] This progress, as in the case of computer storage, has been exponential. For example, Moore's law (which has been confirmed over the past twenty-five years) predicts that technological developments will double the number of elementary logical operators on a microprocessor every eighteen months. This results in an almost linear increase in speed and computational power. To illustrate the rate of such growth, consider that the power of today's largest supercomputers will be found in affordable personal computers within ten years.

Storage

Computerized data are generally held in various kinds of storage *media*. Digitized information can be stored on punched cards, magnetic tape, magnetic disks, optical disks, electronic circuits, smart cards, and biological media. Since the origin of computer technology, storage has evolved toward ever greater capacity, access speed, and reliability as costs have continued to drop.

Progress in computer storage, like that of microprocessors, has been exponential: the space occupied by an 18 MB[2] hard drive on a personal computer in 1983 could hold 10 GB of information in 1993, a thousandfold increase. This rate of growth has lasted for the past thirty years and looks like it will continue until at least 2010 (the foreseeable horizon).

Between 1956 and 1996, the storage capacity of computer hard drives has increased six hundred times, and the density of recorded information by a factor of 720,000. During this same period, the cost of a megabyte of storage dropped from approximately ten thousand dollars to forty cents.[3]

Storage technologies use highly varied materials and processes. Future discoveries in physics and biotechnology, which are actively being pursued in a number of laboratories, may result in advances that are currently unimaginable.

Transmission

Digitized information can be transmitted over almost any imaginable communications channel. Media (disks, diskettes, etc.) can be physically transported by road, rail, boat, or air. But direct, that is, networked, or on-line, connections are obviously faster. Information can be transmitted over conventional phone lines providing it is modulated (using the appropriate analog encoding) when it travels over the telephone network and demodulated (redigitized) when it arrives at a computer or other digital device at the end of the wire. The device that is used to modulate and demodulate digitized information and thus enables communication to take place between computers is called a *modem*. Large, expensive, and slow during the seventies, since the mid-nineties modems have had a transmission capacity that is greater than that of the average user's telephone line. Now common, modems have been miniaturized and are often integrated in computers in the form of cards or printed circuits.

Information can travel unchanged in digital form over copper coaxial cable, fiber-optic cable, radio waves (electromagnetic waves), or, in the case of the telephone network, by means of telecommunications satellite.

Advances in transmission speed (throughput and reliability) depend on several factors. The first is raw transmission capacity. Here the improvements brought about through the introduction of fiber optics have been spectacular. Research is currently under way in a number of labs on the use of "dark fiber," an optical channel in which a single strand, as fine as a human hair, can handle the U.S. telephone system's entire data stream on Mother's Day (when network usage is highest for the year). Hardware that uses this black fiber would provide a thousand times the current broadcast transmission capacity throughout the entire frequency spectrum.

The second is the capacity to compress and decompress information. Currently, sound and video information consume the greatest amount of bandwidth and storage capacity. Some software programs or compression chips can analyze audio and video, simplifying or isolating the most salient parts of the message, which can be several thousand times smaller than in its original digital form. At the other end of the transmission channel, a decompression module regenerates the audio or video from the description it receives, minimizing the loss of information. By compressing and decompressing messages, we transfer part of the problem inherent in transmission (and storage) to the processing unit, which is becoming faster and cheaper.

The third is the advance in the overall architecture of communications systems. Here the greatest progress has been made in the expanded use of packet switching. Such a decentralized architecture, in which every node in the network is "intelligent," was conceived at the end of the fifties as a response to the threat of nuclear war. However, full-scale experiments weren't conducted in the United States until the end of the sixties. With packet switching, messages are cut into small units of equal size, called packets, each of which is provided with a source address, destination address, and position in the complete message. Routers, distributed throughout the network, are able to read and forward this information. The network can be physically heterogeneous (cables, radio waves, satellites, etc.); routers need only to know how to read the packet addresses and speak the same "language" with one another. If information is lost at some point, routers can request that the missing packets be resent. At regular intervals, routers gather information about the status of the network. If there are problems—damage, faults, congestion— packets can take any of a number of different paths, before being reassembled at their ultimate destination. This system is particularly resistant to problems on the network because it is decentralized and its intelligence "distributed." As of 1997 it was operational only on certain specialized networks (notably the Internet backbone), but an international communications standard known as Asynchronous Transfer Mode (ATM), which is consistent with packet switching, has been adopted by the International Telecommunications Union (ITU). In the future, ATM will be used for all telecommunications networks and may lead to the use of broadband digital multimedia communication.

We can get some idea of the progress that has been made in transmission speeds from the following figures. Throughout the seventies, Arpanet (the Internet's predecessor) in the United States could support speeds of 56,000 bits per second. During the eighties, the network cables used to connect American scientists could transmit 1.5 million bits per second. In 1992 data channels on the same network were able to transmit 45 million bits per second (one encyclopedia every minute). Current research foresees the construction of networks with a bandwidth of several hundred billion bits a second (a large library every minute).

Interfaces

An interface is any device that enables interaction between the universe of digitized information and the ordinary world. Input devices capture and digitize information for further processing. Until the seventies, the majority of computers were supplied data in the form of boxes of punched cards. Since then we have expanded the possibility of capturing bodily movement and various physical data *directly:* keyboards to enter text and send instructions, mice for manipulating information on computer screens, touch screens, sound samplers, speech recognition software, scanners for images and text, optical readers (bar codes or other information), automatic sensors (data gloves and suits) for capturing the movements of the body, the eyes, brain waves, and nervous system (used in prostheses), sensors for a wide variety of physical measurements, including heat, humidity, light, weight, and chemical properties.

After they have been stored, processed, and transmitted in digital form, abstract models are made visible, descriptions of images are given shape and color, sounds ring through the air, text can be printed on paper or read on-screen, and automatons controlled by means of actuators. The quality of the display, or *output*, medium is obviously critical for users of computer systems and, to a large extent, conditions their practical and commercial success. Until the seventies, most computers had no monitors. The first computer screens could display only monochrome characters (numbers and letters). Today ultraflat color liquid crystal screens are available, and developments are under way to sell systems that can display images in three dimensions.

The evolution of output interfaces has been accompanied by greater resolution and diversification in the modes of communicating information. Aside from the quality of screen graphics, the quality of printed text and graphics has advanced considerably in less than ten years, erasing the distinction between printed text and manuscript and transforming our relationship to the printed document. Most loud-speakers today play music that is stored (and often produced) in digital format, and considerable progress has been made in the field of speech recognition. There have also been improvements in tactile and pro-prioceptive technologies: force feedback on game controllers, joysticks, and other manual controls; the very sensation of smoothness or rough-ness has helped to enhance the illusion of reality when interacting with virtual worlds.

Two separate research paths are being conducted in parallel for the development of interfaces. One focuses on the immersion of the five senses in increasingly realistic virtual worlds. Virtual reality is used pri-marily for military, industrial, medical, and urban planning applications. Here the human is asked to cross over to the other side of the screen and interact physically with digital models.[4] The other, known as "aug-mented reality,"[5] populates our natural physical environment with sen-sors, cameras, video screens, and interconnected intelligent modules. We no longer communicate with a computer by means of an interface but carry out a multitude of tasks within a "natural" environment that supplies us with the various resources for creativity, information, and communication.

Most communications devices (telephones, televisions, photocop-iers, fax machines, etc.) contain, in one form or another, an interface with the digital world through which they are interconnected. This is equally true of a growing number of machines, measurement devices, so-called mobile objects (personal digital assistants, cellular phones, etc.), personal transportation vehicles, and so forth. The diversification and miniaturization of interfaces, combined with progress in digitization, are converging toward an extension and multiplication of the number of entry points to cyberspace.

Char Davies's Osmosis

September 1995. You are attending an international symposium on electronic art, which is being held in Montreal this year. You made a reservation several days in advance to explore Osmosis, a virtual world created by Char Davies, a Canadian artist. You arrive in the specially equipped room on the first floor of the Museum of Contemporary Art on time. The small room is covered with computers, cables, and electronic devices of all kinds. An assistant helps you up to a platform, above which is suspended an infrared sensor that will record your movements. Slightly terrified, you put on a heavy harness that wraps tightly around your chest. A headpiece, equipped with stereoscopic glasses and headphones, is adjusted. "To go up, inhale. To go down, exhale." The idea of using our breath to move around was suggested to Davies by undersea diving, of which she is an avid practitioner. "To go forward, lean forward. To back up, lean backward. You have twenty minutes. Do you understand? It's not too tight?" Although not exactly comfortable, you indicate with a movement of your head that everything is fine.

You are thrust into interstellar space. A soft, continuous, cosmic music accompanies the gentle gravity, the slow turning motion that leads you toward the luminous planet below, which is your destination. You feel like the fetus that returns to earth at the end of *2001: A Space Odyssey*, by Stanley Kubrick. In slow motion you penetrate the world where you were born, crossing layers of cloudlike computer code, winds of words and phrases, until you finally land in the center of a clearing. From now on, you direct all your movements. Awkwardly at first, then with greater assurance, you experiment with this strange new way of moving. Breathing in sharply, you rise above the clearing. Small animals like fireflies, dancing on the outskirts of the forest, serve as your escort. A pond covered with water lilies and strange aquatic plants glitters before you. The world is soft, organic, dominated by an omnipresent vegetation. Leaning over, you head toward a large tree that appears to serve as the axis of the sacred clearing. Suddenly you come into contact with the tree, penetrating the wood itself. As if you were a sensate molecule, you follow the channel used to carry its sap. Breathing in sharply, you rise inside the tree until you arrive at its topmost branches. Surrounded by capsules of chlorophyll, soft green in color, you enter a leaf, where you watch the complicated dance of photo-synthesis. Leaving the leaf, you hover again above the clearing. You drop down to the pond by exhaling heavily and again cross a swarm of fireflies (sprites?) that give off a strange sound like bells ringing in the distance. Turning your head, you watch as they fly toward the forest, accompanied by the lingering echoes of the fading sound of celestial bells. Now you are nearly on top of the surface of the pond, where you stop to watch the play of reflected light. As you break through the water's surface, a fish, undulating gracefully, welcomes you to the aquatic world.

After visiting the pond, you cross the forest, the mineral world, and enter a bizarre space, crisscrossed by lines of writing. By breathing in and out, by moving your torso, you make out a few sentences of philosophy: the world of human discourse encompasses the natural environment. Finally you reach the world of data, populated with lines of code. You assume you'll have time to come back to these different worlds. But you are pulled upward and gently but firmly led outside the planet Osmosis. Life inside this universe knows only one time. As the planet you experienced for no more than a brief instant disappears from view within the background of interstellar space, you regret that you didn't use your time more wisely. Where will your next incarnation take place?

The principles that guided the design of Osmosis are anti-thetical to those that govern video games. You can't use your hands. The posture typical of prehension, manipulation, or combat is diminished. On the contrary, to move in this vegetal and meditative world, you are forced to concentrate on your breathing and kinesthetic sensations. You must be in *osmosis* with this virtual reality before you can understand it. Sudden or brusque movements are inefficient. Gentle movements and a contemplative attitude are "rewarded." Instead of bright colors, the worlds of the tree, pond, clearing, and forest offer us a subtle mosaic of greens and browns, which evoke vegetable dyes more than they do the garish technology of synthetic images. Osmosis represents the escape of the virtual arts from their original matrix of "realistic" and geometric simulation. The work serves as an overwhelming rebuttal to those who see virtuality as nothing more than a "macho Western project for conquering nature and manipulating the world." Here the virtual is explicitly designed for contemplation, self-awareness, respect for nature, a form of knowledge and relation to the world that is osmotic.

Programming

Cyberspace doesn't consist only of hardware, data, and human beings; it is also populated with strange beings, half text, half machine, half actor, half script, known as programs. A program, or software application, is a well-organized series of coded instructions designed to accomplish a particular task on one or more processors. Using the circuitry they control, programs can interpret data, modify information, change other programs, activate computers, networks, and physical machinery, travel, reproduce.

Programs are written in programming languages, which are specially encoded using the instructions that are understood by computer processors. There are a large number of programming languages in existence, each of which is more or less specialized in a certain task. Since the beginning of the computer age, engineers, mathematicians, and linguists have been working to make programming languages more and more like natural language. We distinguish between the esoteric programming languages that are very close to the physical structure of the computer (machine languages, assembly language) and "advanced" languages, which are less dependent on the structure of the hardware and closer to English. These languages include FORTRAN, Lisp, Pascal, Prolog, C, and others. Today some so-called fourth-generation languages can be used to design programs by drawing diagrams and manipulating icons on-screen. Companies supply software building blocks that are ready for assembly. In this way, the programmer spends less time coding and more time designing the overall software architecture. Authoring languages enable nonspecialists to perform simple programming tasks and to design multimedia databases and educational software.

Software

Software applications are used by computers to provide specific services to users. I'll give a few canonical examples. Some software programs can automatically calculate employee payrolls for a company; others can prepare customer invoices, manage inventory, and operate machines in real time based on information supplied by sensors. Expert systems can determine the causes of a failure or supply financial advice. As the name indicates, word processors are used to write, modify, and organize texts. Spreadsheets display columns of numbers, can be used for accounting, and help us make budget decisions. A database manager can be used to enter data in files and retrieve pertinent information based on various input keys, presenting the information to the user from a given point of view depending on need. Drawing software can be used to produce flawless diagrams almost effortlessly. Communications software enables us to send messages and access information stored on other computers. Application software is becoming increasingly open to the personalization of features, without requiring that users learn programming.

Operating systems are programs that manage computer resources (memory, input, output, etc.) and organize the interface between hardware and application software. This application software is not therefore in direct contact with the hardware. Because of this, the same software application can run on different hardware platforms, providing they use the same operating system.

Although not all data are programs, all programs can be considered data: they can be input, archived, and read by computers. And in particular, they can themselves be calculated, translated, modified, or simulated by other programs. Because a program can replace a collection of data that is being translated or processed by another program, successive software layers can exist between the hardware and the end user. The end user communicates directly with the topmost layer and doesn't need to understand the complexity underlying the application he or she is manipulating or the heterogeneity of the network on which he or she is transferring data. In general, the greater the number of software layers, the more the network is "transparent," and the easier it is for us to accomplish human tasks.

From the Computer to Cyberspace

Today we can freely navigate between software and hardware that was formerly incompatible. Through the adoption of software and hardware standards, the general trend has been toward the development of autonomous virtual spaces for work and communication, increasingly independent of physical media. There has also been an increasing use of standards for describing the structure of text (Standard Generalized Markup Language, or SGML) or multimedia (Hypertext Markup Language, or HTML, HiTime)[6] documents, which enable us to retain information intact in spite of changes in software and hardware. The VRML standard can be used to explore interactive three-dimensional images on the World Wide Web from any machine connected to the network.[7] The increasing use of the VRML standard presages the interconnection of virtual worlds that can be accessed on the Internet and expands the boundaries of a cyberspace that is similar to an immense heterogeneous virtual metaworld undergoing continuous transformation and containing all other virtual worlds.

Long polarized by the "machine," balkanized by software, contemporary information technology—software and hardware—deconstructs the computer and promotes a navigable and transparent communications space based on information. A computer is a particular assemblage of units for processing, transmission, and storage, and interfaces for information input and output. Yet different brands of computer can be assembled from nearly identical components, and computers from the same company can contain parts from very different sources. Moreover, hardware components (sensors, memory, processors, etc.) can be found outside actual computers: smart cards, automatic distributors, robots, motors, household appliances, automobiles, photocopiers, facsimile machines, video cameras, telephones, radios, televisions, and even the nodes of the communications network ... anywhere digital information is automatically processed. A computer connected to cyberspace can use the storage capacity and processing power of other computers on the network (which do the same), as well as remote devices for capturing and displaying information. All computer functions are distributable and, to an ever increasing extent, distributed. The computer is no longer a center but a node, a terminal, a component of the universal computing network. Its atomized functions impregnate every element of the technocosm. Eventually there will be only a single computer, but it will be impossible to locate its boundaries or determine its contour. Its center will be everywhere, and its circumference nowhere. This hypertext computer will be dispersed, living, pullulating, incomplete: cyberspace itself.

CHAPTER 3

Digital Technology and the Virtualization of Information

The Golden Calf

Not far from the basilica of Saint-Denis, which contains the funerary monuments of the former kings of France, Artifices, an exposition devoted to digital technology, is held every two years. In November 1996 the feature artist was Jeffrey Shaw, a virtual arts pioneer and director of an important German institute devoted to "new media."

When you enter the exhibit, the first thing you see is an installation of the golden calf. In the center of the first room, a pedestal, obviously designed to hold a statue, sits empty. The statue is missing. A flat screen rests on a table next to the pedestal. Upon closer examination, you realize that this liquid crystal screen serves as a "window" on the room around you: by turning it toward the walls or ceiling, you see a digital image of the walls or ceiling. Turning it to the entrance, you see a digital model of the doorway appear on-screen. When you turn the screen toward the pedestal, you discover a beautifully sculpted and shining statue of the golden calf. The sculpture is visible only through the screen; its "existence" is virtual. Walking around the pedestal, still holding the screen in the direction of the emptiness above it, you can admire every aspect of the statue. As you approach, it gets larger; as you withdraw, it grows smaller.

If you move the screen high enough above the pedestal, you can look inside the statue and see the secret within: the inside is empty. The calf exists only on its outside surface; there is no inside, no interior.

What does the installation mean? It is primarily a criticism: the virtual is the new golden calf, the new idol of our time. But it is also classical. The work enables us to concretely experience the essence of any idol: an entity that isn't really there, an appearance without consistency, without interiority. Here it is not so much the absence of material plenitude that is on display as the nothingness of a living, subjective presence and interiority. The idol has no existence other than that assigned to it by those who worship it. The relationship to the idol is enacted through the mechanics of the installation, since the golden calf is visible only through the visitor's actions.

On a level where aesthetic problems connect with spiritual doubt, Jeffrey Shaw's installation questions the notion of representation. The golden calf obviously refers to the second of the Ten Commandments, which prohibits not only idolatry but the manufacture of images and statues "whose forms resemble those found in the sky, on earth, or in the waters." Has Jeffrey Shaw sculpted a statue or created an image? Is his golden calf a representation? There's nothing on the pedestal. The life and sentient interiority of that which flies in the air or runs on the ground have not been captured in a dead form. It isn't a calf, worshiped in a material we hold precious, that the installation presents to us, *but the very process of representation*. While in some ultimate sense there is only nothingness, the visitor's mental and sensorimotor activity causes an image to appear, which, once it has been explored thoroughly, ends by revealing its emptiness.

This chapter is devoted to the new kinds of messages that proliferate on computers and networks: hypertexts, hyperdocuments, interactive simulations, virtual worlds. I will try to show that virtuality, in its broadest sense, constitutes the distinctive feature of the new face of information. Because digitization is the technical foundation of virtuality, an explanation of its principles and functions follows my introduction to the concept of the virtual, which begins the chapter.

The Virtual

Globalization, the first fundamental aspect of cyberculture, has propagated the copresence and interaction of points anywhere in the physical, social, and information space. In this sense, it complements a second fundamental aspect—virtualization.[1]

The word "virtual" has at least three meanings: a technical meaning associated with information technology, a contemporary meaning, and a philosophical meaning.[2] Our fascination with "virtual reality" is based in large part on a confusion of these three meanings. In its philosophical sense, the virtual is *that which exists potentially rather than actually*, the field of forces and problems that is resolved through *actualization*. The virtual precedes its effective or formal concretization (the tree is *virtually* present in its seed). Philosophically speaking, the virtual is obviously a very important dimension of reality. But as it is currently employed, the word "virtual" often signifies unreality, "reality" here implying some material embodiment, a tangible presence. The expression "virtual reality" sounds like an oxymoron, some mysterious sleight of hand. We generally assume that something is either real or virtual, and that it cannot possess both qualities at once. For a philosopher, however, the virtual stands in opposition not to the real but to the actual, virtuality and actuality being nothing more than two different modes of reality. Although the seed's essence is the production of a tree, the tree's virtuality is quite real (without yet being actual).

Any entity is virtual if it is "deterritorialized," capable of engendering several concrete manifestations at different times and places, without being attached to any particular place or time. To use an illustration outside the world of technology for a moment, a *word* is a virtual entity. The word "tree" continues to be used in different places, at different times. Every time the word is pronounced, it is "actualized."

But the word itself, the word that is pronounced or actualized, exists nowhere and is not attached to any precise moment in time (regardless of how long it may have existed).

Although we are unable to assign it any spatial and temporal co-ordinates, the virtual is nonetheless real. Words exist. The virtual exists without being anywhere. It is also worth noting that actualizations of the same virtual entity can be vastly different from one another; the actual is never completely predetermined by the virtual. Thus from both an acoustic as well as a semantic point of view, no given actualization of a word is exactly like another. Unforeseeable pronunciations (new voices) and meanings (the invention of new sentences) can always arise. The virtual is an infinite source of actualization.

Cyberculture is linked to the virtual directly and indirectly. Directly, the digitization of information can be assimilated to virtual-ization. The computer codes written on diskettes or computer hard drives—invisible, easily copied or transferred from one node of the network to another—are quasi virtual, since they are nearly indepen-dent of any determined spatiotemporal coordinates. Within digital net-works, information is obviously physically present somewhere, on a given medium, but it is also virtually present at each point of the network where it is requested.

Digital information (0s and 1s) can also be qualified as virtual to the extent that it is inaccessible to humans. We can only directly inter-act with its actualization, through some form of display. Computer codes, which are illegible, are actualized as readable text, visible images on computer screens or paper, audible sounds in the atmosphere.

Images observed during our exploration of some "virtual reality" are not stored as such in a computer memory. More likely than not, they are calculated in real time (at the moment of our request) based on an informational matrix containing the description of the virtual world. The computer synthesizes the image from the data (constant) in this matrix and other information (variable) regarding the "position" of the explorer and his or her previous actions. A virtual world—considered as a group of digital codes—is an *image potential*, and any given view, displayed during the course of our immersion in this virtual world, actualizes this potential within a particular context. This dialectic of potential, calculation and contextual display, characterizes the majority of documents or information sets stored on digital media.

Indirectly, the development of interactive digital networks promotes other forms of virtualization in addition to those associated with information. Along with digital technology, communications have for years been engaged in a movement toward virtualization through technologies such as writing, sound and image recording, radio, television, and telephony. Cyberspace encourages a relation that is nearly independent of geographic location (telecommunications, telepresence) and temporal coincidence (asynchronous communications). This is not a completely novel idea, since the telephone has already accustomed us to interactive telecommunications. In the post office (or writing in general) we have a tradition, dating back many years, of reciprocal communication that is both asynchronous and remote. Yet it is only the specific technological characteristics of cyberspace that enable members of a human group (regardless of their number) to cooperate with one another, to nourish and consult a shared memory; and they can do so almost in real time regardless of geographic distribution and differences in time zones. This leads directly to the virtualization of organizations, which, by using the tools of cyberculture, are becoming less and less dependent on specific locations, fixed hours, and long-term planning. Similarly, through cyberspace, economic and financial transactions further accentuate the virtuality that has characterized them since the invention of money and banking.

In short, the extension of cyberspace accompanies and accelerates a general virtualization of the economy and society. In the case of substances and objects, we examine the processes that produced them. In the case of geographic territories, we take a step backward to the mobile networks that enhance and describe them. In the case of processes and networks, we move toward the skills and patterns that control them— an ever increasing level of virtualization. The media of collective intelligence in cyberspace are proliferating, helping to synergize the skills of its participants. From design to strategy, patterns are generated out of the simulations and data supplied by the digital universe.

Ubiquitous information, interconnected interactive documents, reciprocal asynchronous communication within and among groups—the virtualizing and deterritorializing character of cyberspace has made it the vector of an open universal. At the same time, the extension of a new universal space expands the field of action for the processes of virtualization.

Digital Technology

Digitizing information consists in translating it into numbers. Nearly all information can be encoded in this way. For example, if we assign a number to each letter of the alphabet, any text can be transformed into a sequence of numbers.

An image can be decomposed into dots or pixels (picture elements). Each dot can be described by two numbers specifying its coordinates on the plane and by three numbers reflecting the intensity of each of its color components (red, blue, and green). In this way, any image or sequence of images can be translated into a sequence of numbers.

A sound can also be digitized by being sampled, which involves measuring it at regular intervals (more than sixty thousand times a second when recording high frequencies). Each sample can be encoded by a number that describes the sound signal at the time of measurement. Any audio or musical sequence can therefore be represented by a series of numbers.

Images and sounds can also be digitized, not only dot by dot or sample by sample, but more efficiently using a description of the overall structure of the graphic or auditory message. Sinusoidal functions are used for sound, and functions that generate geometric figures for images.

In general, any kind of information or message, providing it can be described or measured, can be digitized.[3] Moreover, all numbers can be expressed in binary language, consisting of 1s and 0s. This means that all information can ultimately be represented in binary form, which has several advantages.

There are large numbers of devices available that can record and transmit numbers encoded in binary format. These binary numbers can be represented physically in either of two states (open or closed, negative or positive, etc.), which circulate along electrical wires and circuits, polarize magnetic tape, are translated into flashes in optical fibers and micropits on optical disks, and can be represented in molecular structures.

Digitally encoded information can also be transmitted and copied almost indefinitely without the loss of information, since the original message can be regenerated in spite of any degradation caused by the transmission path (telephone or radio waves, for example) or process of duplication. This is obviously not true for images and sound recorded by analog means, which degrade every time they are copied or transmitted. Information encoded using analog means establishes a proportional relationship between a given parameter of the information being translated and a given parameter of the translated information. For example, sound volume can be encoded as the intensity of an electrical signal (or the depth of a groove on a vinyl record): the higher the volume, the greater the intensity of the electrical signal (or the deeper the groove). Analog information is represented as a continuous series of values. Digital encoding, on the other hand, uses only two values, which are clearly differentiated. This makes it far easier to regenerate noise-corrupted information using techniques that control the integrity of the message.

Binary encoded numbers can also be used for arithmetic and logic calculations in specialized electronic circuits. Even though we speak of the "immaterial" or "virtual" nature of digital space, it is important to remember that digital processing is always based on simple physical operations on physical representations of 0 and 1: deletion, substitution, sorting, ordering, pointers to a given storage area or transmission channel. Once processed, binary encoded information can be translated (automatically) in the reverse direction and can be represented as readable texts, visible images, audible sounds, tactile or proprioceptive sensations, or even the movements of a robot or mechanical device. Why are ever increasing amounts of information not only being digitized but directly produced in digital form? The primary reason is that digitization provides an opportunity to process complex information efficiently, which would be impossible using other methods.

The Automatic Processing of Information

Digitized information can be processed automatically, at an almost absolute level of detail, with extreme speed and in large volumes. No other process can achieve all these qualities at the same time. Digitization can be used to control information "bit by bit," binary number by binary number, at the speed of a computer.

I'll start by giving a simple example. Take a three-hundred-page novel that has been digitized. Using a word processor, I can have my computer replace all the occurrences of "Smith" by "Jones." The computer will carry out this order in a few seconds. On my hard drive, the permanent magnetic storage area in my computer, where information is stored in binary form, all the names are changed almost immediately. If the text had been printed on paper, the same operation would have required considerably more time. I can also reverse the order of two chapters and change the pagination in a matter of seconds. I can change the typeface, whereas the same operation using lead type would require that the text be reset.

How about sound? Once a violin passage, for example, has been sampled, the appropriate software can slow down or speed up the tempo without modifying the frequency of the sounds (flats and sharps). We can also isolate the timbre of the instrument and have it play a different melody. Using the same musical passage, we can calculate (and play back) the continuous transition of the timbre of the violin to that of a piano. Once again, this type of manipulation is almost impossible to obtain rapidly and automatically without digital processing.

Assume a film has been digitized. Specialized software can be used to automatically, and almost instantaneously, change the color of a flower or dress in all frames of the film. In a digitized photograph, the size of an object can be reduced by 17 percent, for example, without altering its shape. If the representation is in three dimensions, we can automatically calculate a new perspective when the point of view of the scene is rotated nine degrees to the left. All these operations can be performed nearly instantaneously. The reason for this is that information is encoded in the form of numbers that can be manipulated with ease: numbers can be calculated, and calculation is one of the things computers can do very quickly.

It's possible not only to process information but to generate it. Some musical synthesizers can produce sounds based on samples of natural sounds; others can produce sounds solely on the basis of physical models of the sound to be produced or even a mathematical description of the vibrations of the instrument being imitated. Similarly, some digital films are created not by processing a hand-drawn image or one photographed with a camera but from geometric models of volumes and the laws for light refraction, mathematical functions describing the movements of the characters or a virtual camera, and so on. Synthesis programs, incorporating formal models of the objects that will be simulated, use computers to calculate images and sounds.

Dematerialization or Virtualization?

Is digitization a "dematerialization" of information? To better understand the question, let's look at an example. Take a photograph of a blossoming cherry tree, obtained by optically capturing the image and chemically treating it in a silver chloride bath. Then digitize the photograph using a scanner. It now exists as a sequence of numbers on the hard drive in our computer. In one sense, the photograph has been "dematerialized," since the sequence of numbers is a very accurate *description* of the photograph of the blossoming cherry tree and no longer a two-dimensional image. Nonetheless, the description itself can't exist without a physical medium. It occupies a specific area of space; it requires some form of recording hardware, a device that costs money and has weight, a certain amount of physical energy to be recorded and regenerated. We can then translate this encoded description on the computer into a visible image on a number of different media: an image on a screen, a paper printout, or some other process.[4] The digital encoding of the blossoming cherry tree is not, strictly speaking, "immaterial" but simply occupies less space and weighs less than a paper photograph. It requires less energy to modify or manipulate the digital image than the silver image. Fluid and volatile, the digital recording occupies a unique position in the succession of images: implicit in their visible manifestation, it is neither unreal nor immaterial, but *virtual*.

From a single negative, a conventional photograph can be enlarged, retouched, developed, and reproduced in as many copies as we wish. What do we gain through digitization? Where is the qualitative difference? Not only can the digitized image be modified more easily and quickly, but it does not require methods of mass reproduction to be made visible. For example, using the appropriate software, the cherry tree can be represented with or without its foliage, depending on the season, or at different sizes, depending on where in the garden we wish to put it, or with different color flowers.

Let's take a last look at that cherry tree in blossom. It can be drawn, photographed, or digitized from a conventional photograph and retouched by the computer; it can even be synthesized entirely by a computer program. If we consider the computer as a tool for processing or producing images, it becomes merely another instrument, whose efficiency and degrees of freedom are greater than those of a brush or camera. The ontological status or aesthetic property of the image as such, although produced by computer, is not fundamentally different from any other type of image. Yet if we no longer consider a single image (or a single film) but all the different images (or films) that can automatically be produced by a computer from the same digital engram, we penetrate a new universe for generating signs. Starting with an initial data store, a collection of descriptions or models, a program can calculate an indefinite number of different visible, audible, or tangible manifestations based on the current situation or the users' request. The computer is no longer just a tool for producing texts, sounds, or images, but a means of *virtualizing information*.

Hyperdocuments

A CD-ROM (compact disc read-only memory) or CD-I (compact disc interactive) is a medium for digital information read by a laser. It can contain sound, text, and images (stationary or animated) for display on computer screens in the case of a CD-ROM or on a television (equipped with a special reader) in the case of a CD-I. When viewing a CD-ROM, we "navigate" the information on it, move from one screen page or animation to another by indicating the topics that interest us or the text links we wish to follow. Navigation takes place by clicking with a mouse on an icon on-screen, by pressing a key, by manipulating a remote control, or by activating a joystick. Encyclopedias, art history, music, and games—CDs are the best-known hyperdocuments at this time (1997). The CD-ROM (which can hold the entire text of a thirty-volume encyclopedia) will soon be replaced by DVD (digital video disc) with six times as much storage capacity, enough to hold a full-screen video film.

Using the broadest possible meaning for "text" (one that doesn't exclude sound or images), hyperdocuments could also be called *hypertexts*. The simplest way to approach hypertext is to describe it as a text structured as a network. The hypertext consists of nodes (elements of information, paragraphs, pages, images, musical sequences, etc.) and links among these nodes, references, notes, pointers, and buttons for moving from one place to another within the document.

A novel is generally read from front to back, a film from the first frame to the last. But how do we read an encyclopedia? We can start by looking at the index or thesaurus, which will refer us to one or more articles. At the end of the article is a reference to other articles on related subjects. Each of us navigates the encyclopedia, with the topics that interest us serving as a guide, and beats a unique path through the body of information it contains, using whatever navigational tools are available—dictionaries, lexicons, indexes, thesauri, atlases, tables of numbers and tables of contents, which are themselves small hypertexts. Keeping to our definition of a "networked text" or documentary network, a library can be considered a hypertext as well. In this case the link between volumes is provided by references, footnotes, citations, and bibliographies. Card files and catalogs serve as the tools of global navigation in the library.

Digital media, however, introduce a considerable difference compared with precomputer hypertexts: searching an index, using navigational instruments, or moving from one node to another all take place with great speed, on the order of a few seconds. Digitization can also be used to combine and mix sound, image, and text on the same medium. From this point of view, digital hypertext could be defined as multimodal information arranged in a network for rapid and "intuitive" navigation. Compared to earlier technologies for reading, digitization introduces a small Copernican revolution of its own: we no longer have a navigator who moves physically through the hypertext, turning pages, shifting heavy volumes, pacing up and down before stacks in a library, but a mobile, kaleidoscopic text, which offers its facets to the reader, turns, folds, and unfolds before our eyes.

A new art of printing and documentation is being invented, which attempts to take advantage of the speed at which we can navigate masses of information, condensed into ever smaller volumes.

The contemporary tendency to hypertextualize documents could also be defined as a tendency toward indistinctness, toward a blending of the functions of reading and writing. Let's examine the situation from the reader's point of view. If we define a hypertext as a space of possible readings, a text would appear to be a particular reading of a hypertext. Here the navigator helps *write* the text he or she reads. It's as if the author of a hypertext had built a matrix of potential texts, the navigator's role being to realize certain texts by manipulating the way the nodes are combined. The hypertext brings about the virtualization of the text.

The navigator can become an author in a more profound sense by traversing a preestablished network—by participating in structuring a hypertext. The navigator no longer merely follows preexisting links but creates new links, those that have meaning for him or her and which the creator of the hyperdocument has overlooked. Systems can also record these movements and strengthen (by making them more visible, for example) or weaken links based on the way they are traversed by the community of navigators.

Finally, readers can not only modify these links but also add or modify nodes (texts, images, etc.), connect one hyperdocument to another, creating a single document out of two separate hypertexts, or possibly create hypertext links among a multitude of documents. This technique is currently being developed on the Internet, especially through the World Wide Web. In the two last forms of navigation, hyperdocuments are no longer stored on CDs but are accessible on-line to a community of persons. When the system for the real-time display of the hypertext (its dynamic mapping) is well designed, or when navigation can take place naturally and intuitively, open hyperdocuments accessible over a computer network become powerful instruments of *collective reading-writing*.

From the author's point of view, the large masses of information collected in hyperdocuments come from various sources. The selection and addition of this information to the network could be considered one of their possible "readings." The author, or more often a team of writers, makes use of hardware, software, and elements of existing interfaces for constructing the hyperdocument. This results in a specific navigational path among the available information, hardware, and software. The editorialized hyperdocument is itself a path inside a much larger and more obscure hyperdocument.

Writing and reading have exchanged roles. The person who participates in structuring a hypertext, in tracing the outlines of the possible folds of meaning it contains, is already a reader. Similarly, the person who actualizes a path or manifests a given aspect of the document archive, helps write it, momentarily completes an interminable text. The disjunctions and cross-references, the original paths of meaning the reader invents, can be incorporated in the very structure of the corpus. With the hypertext, every reader is a potential writer.

Actuality of the Virtual

The scene shows a young girl of flesh and blood blowing a kind of whistle. Reaction shot: on-screen, a computer-generated image of the seeds of a dandelion separating from the stalk and floating away on the breeze. The young girl continues to blow into the whistle. Reaction shot: on the computer screen, a computer-generated cloud rises gently in the currents of virtual air modeled by Edmont Couchot, Michel Bret, and Marie-Hélène Tramus.

The CD *Actuality of the Virtual*, published in *Revue virtuelle* by the Pompidou Center, provides an overview of the current status of the digital arts, interactivity, and networking. It contains all twenty-five conferences sponsored by the review between 1992 and 1996, along with 155 excerpts from the artworks and interactive CDs that were presented to the public.

A sequence by Karl Sims is animated with "artificial life" programs, simulating the growth, genetic mutation, and interaction of imaginary populations. Fibrillation, expansion, color, brightness, and the emergence of unanticipated forms animate the field of a strange image that is never the same, that reacts in real time to the tactile stimulation of the interactive viewer.

Anne-Marie Duguet explains the profound affinities that unite the virtual arts to video art and the work on "installations" that has been done over the past fifty years. Some people claim that digital art is new because the technology is new; others claim it is nothing more than mystification. The word "mystification" turns red whenever the cursor passes over it. When you click on this scarlet letter, it turns blue. There then appears at the top of the screen the titles of those paragraphs from other conferences where the word occurs. Click on the name in a paragraph by Jean-Baptiste Barrière of IRCAM. His text is before your eyes. Do virtual works look toward a new "total art," or are digital artists simply producing better video games? On the left side of the screen, a table of contents allows you to move rapidly through the text. On the far left, you can make out the edge of an interactive graphic illustrating the text, which you can drag to the center of the screen.

Sitting in front of the digital face programmed by Keith Waters, you wonder how to react. Gaining courage, you decide to move the small hand that has replaced the cursor on the computer-generated image. The face suddenly wrinkles its eyebrows and tries desperately, all the while making frantic gestures, to avoid contact with the unwanted cursor. Its reactions vary depending on whether you "tickle" its eyes, nose, or mouth.

Returning to Jean-Baptiste Barrière's text, you follow the hypertext link to a conference by Alain Le Diberder on video games, which is also abundantly illustrated. Reading the text, you click on "Glossary," which displays in ultramarine blue all the words in the paragraph along with a highly informative formal definition. The definition appears by simply clicking on a word. You move from the Le Diberder conference to one by Florian Roetzer, which explains how video games parallel the new cognitive skills required in new forms of work: speed, the ability to manipulate complex models, the discovery of nonexplicit rules through exploration, and so forth.

Illustrations on the theme of interactivity are found nearly everywhere. An image changes depending on where the viewer looks (an image of the viewer's face is captured by a hidden camera and analyzed by a computer program). You explore an environment by holding a large ball representing the eye at arm's length. The device enables you to "see" as if your eye were attached to the tip of your hand. You alter the movements of an artificial swarm of butterflies by moving the beam of a real flashlight on the surface of the projected image.

Following the hypertext links, you arrive at a text by David Le Breton, who claims that virtual technologies will eliminate or reify the body, that these are merely the continuation of the old Judeo-Christian attempt of the Western world to dominate nature. Obviously Le Breton hasn't spent much time in Osmosis. At the bottom of the screen, the faces of the conference speakers appear for a few seconds, only to be replaced by the faces of different speakers. Intrigued, you click on the face of Derrick de Kerckove, who explains in his own voice that virtual and telepresence technologies extend and enhance the sense of touch—contradicting what you have just read. The CD is organized in such a way that it simulates a fictional conversation among the conference participants, each of whom provides examples to support his or her point of view. The navigator controls the rhythm and direction of this virtual conversation, frees the manifold discourses engraved on the disc. To prevent viewers from going in circles, links that have already been followed do not appear again during the same session.

The image of a man is gradually and invisibly morphed into the image of a monkey. The digital is the realm of metamorphoses.

Voxelmann, the virtual atlas of the anatomy, provides an unlimited number of views of a digital model of the body. I marvel at the incredible complexity of the sinus passages.

Five into One, Matt Mullican's virtual city, installs a philosophical concept, an abstract cosmology within a three-dimensional space. Does the virtual image prefigure some sensible embodiment of the world of ideas?[1]

The Japanese maple modeled by the Centre International de Recherche sur l'Agriculture et le Développement (CIRAD) is first shown during the winter, with a gust of wind on the sound track. Buds begin to appear, the branches turn a soft green, the tree fills with birds. The tree's foliage grows more abundant, denser, a darker green. We hear frogs croaking in the background, typical of summer nights. Then the leaves begin to turn yellow, rust, fall to the ground. Winter begins again. The simple poetry of the seasons, the unforgettable contraction of time, is evoked by the computer-generated image.

Multimedia or Unimedia?

Because the word "multimedia" has been subject to so much confusion, I'd like to define a few key terms in the field of information and communication.

The *medium* is the substrate or vehicle of the message. Printed matter, radio, television, film, and the Internet are media.

The reception of a message can involve several *perceptive modalities*. With print media, the sense of sight is primary, and the sense of touch secondary. Ever since the birth of the talkie, film has involved two senses: sight and sound. Virtual realities can incorporate sight, sound, touch, and kinesthesia (the internal sense of bodily movement).

The same perceptive modality can accommodate the reception of different *representational types*. For example, printed matter (which involves only sight) carries text and images. The audio disc (which involves our sense of hearing) can be used to transmit both speech and music.

Analog and digital *encoding* refers to the fundamental method used to record and transmit information. The vinyl record encodes sound analogically, and the audio CD encodes it digitally. Radio, television, cinema, and photography can be analog or digital.

The *information system* refers to the structure of the message or the way in which elements of information are related to one another. The message can be linear (as with ordinary music, the novel, or film) or networked. Digitally encoded hyperdocuments did not introduce networked structure. As I mentioned earlier, dictionaries (in which each word serves as an implicit reference to other words and which we do not read from beginning to end), encyclopedias (containing an index, thesaurus, and multiple cross-references), and libraries (card files and cross-references from book to book) already possess a reticulated structure. Cyberspace did, however, introduce two original information systems that didn't exist in earlier media: the virtual world and information streams. The *virtual world* arranges information in a continuous space—not in a network—based on the position of the user or his or her representative in this world (known as immersion). In this sense, a video game is a virtual world. An *information stream* is a sequence of continuously changing data dispersed among interconnected storage spaces and channels that can be traversed, filtered, and presented to the cybernaut based on his or her instructions by means of software agents, dynamic data mapping systems, or other navigational aids. Through the use of improved technical media, both virtual worlds and information streams tend to reproduce an "unmediated" relationship to information on a large scale. The concept of an information system is, in principle, independent of the medium, the perceptive modality involved, or the type of representation carried by the message.

Finally, a *communications system* designates the relation among participants who are communicating. There are three major types of communications system: one-to-many, one-to-one, and many-to-many. The press, radio, and television are one-to-many systems: a transmitter sends messages to a large number of passive receivers dispersed over a wide area. The post office and telephone organize reciprocal relations between interlocutors, but only through one-to-one or point-to-point contacts. Cyberspace implements an original communications system because it enables communities gradually and cooperatively to build a shared context (many-to-many). On an Internet forum, for example, participants send messages that can be read by all other members of the community and to which any of them may respond. Uninterrupted communication sediments a collective memory that emerges from communication among the participants. Multiparticipant virtual worlds, systems for learning or collaboration, or, on a much larger scale, the World Wide Web, can be considered many-to-many communications systems. Once again, the communications system is independent of the meanings implied by the receiver or the way in which information is represented. I emphasize this point because it is these *new information (virtual worlds, information streams) and communications systems (many-to-many communication) that are most responsible for our cultural mutations* and not the fact that we can mix text, graphics, and sound, which the rather amorphous concept of "multimedia" implies.

In principle, the term "multimedia" signifies the use of several media or several vehicles of communication. Unfortunately, it is rarely used in this sense. Today the word generally refers to the two outstanding tendencies of contemporary communications systems: multimodality and digital integration.

In the first place, computers process not only numerical data or text (which was the situation until the 1970s) but also, and increasingly, images and sound. Linguistically, it would be much more accurate to speak of multimodal information or messages, since they incorporate several sensory modes (sight, hearing, touch, proprioceptive sensation). The term "multimedia" used to designate CD-ROMs is, in my opinion, misleading. If we want the term to mean "multimodal," then it does not sufficiently describe the specificity of these new media. Encyclopedias, even certain children's books and illustrated brochures accompanied by tapes (like those used to learn a language), are already multimodal (text, image, sound, touch), or even multimedia. Strictly speaking, we should define CD-ROMs and CD-Is as digital interactive multimodal documents, or simply as hyperdocuments.

In the second place, the word "multimedia" refers to the general movement of digitization, which involves a range of media at various stages of completion, including information technology (by definition), telephony (in progress), music (already completed), publishing (partially realized with CD-ROM and interactive CDs, on-line books and newspapers), radio, photography (in progress), cinema, and television. Although digitization is moving ahead at high speed, the integration of these media remains a long-term trend. It's possible, for example, that television, even when digitized and more "interactive" than it is today, will remain a relatively distinct medium.

The term "multimedia" is correctly used when, for example, the release of a film occurs simultaneously with the sale of a video game, the distribution of a TV series, T-shirts, toys, and so on. Such a scenario is indeed representative of a "multimedia strategy." But if we want to accurately represent the confluence of separate media within the same integrated digital network, we should use the term "unimedia." The term "multimedia" is prone to error, for it appears to indicate a variety of media or channels, whereas the actual trend has, on the contrary, been one of interconnection and integration.

Whenever we hear or read the word "multimedia" in a context in which it doesn't refer to a particular type of media (see the discussion on

TABLE 1 The Different Dimensions of Communication

Communication Type	Definition	Example
Media	Substrate for information and communication	Printed matter, film, radio, television, telephone, CD-ROM, the Internet (computers plus telecommunications)
Perceptive modality	Meaning implied by the reception of information	Sight, sound, touch, smell, taste, kinesthesia
Language	Type of representation	Spoken languages, music, photographs, drawings, animations, symbols, dance
Encoding	Method of recording and transmitting information	Analog, digital
Information system	Relationships among elements of information	Linearly structured messages (conventional text, music, film), networked messages (dictionaries, hyperdocuments), virtual worlds (information *is* the continuous space, the explorer or representative is immersed in this space), information streams
Communications system	Relationship among participants in a communication	One-to-many system in a hub-and-spoke arrangement (press, radio, television), one-to-one networked system (post office, telephone), many-to-many system in space (mailing lists and newsgroups, systems for learning or collaboration, multiparticipant virtual worlds, WWW)

CD-ROM) or processing, we should assume that the author is referring to a *multimodal unimedia approach*, the gradual constitution of an integrated, digital, and interactive communications infrastructure.

Finally, the word "multimedia" is particularly unfortunate when used to designate the emergence of a new medium, since it draws attention to the forms of representation (text, image, sound) or their medium, whereas the principal novelty lies in the creation of interactive, community-based information (networks, data streams, virtual worlds) and communications systems, that is, ultimately, a relational mode among persons, the quality of the social bond.

Simulation

Before flying a plane for the first time, it's common practice to perform tests to determine how the wings react to wind, air pressure, and atmospheric turbulence. For obvious reasons of cost, it would even be preferable to have some idea of the wing's resistance before construction of a prototype. To do so we can construct a reduced-scale model of the airplane and subject it to violent blasts of wind in a wind tunnel. This was standard procedure for years. As the power of computers has increased and their cost decreased, it is now faster and less expensive to provide the computer with a description of the aircraft and the wind and ask it to calculate a description of the effect of the wind on the airfoil. In this case the computer has *simulated* the aircraft's resistance to the air. For the computer to provide a correct response, the descriptions supplied, for both the aircraft and the wind, must be meticulous, precise, and consistent. Such careful descriptions of objects or phenomena used for simulation are known as *models*.

The result of the simulation can be supplied in the form of a list of numbers, indicating, for example, the maximum pressure on every square centimeter of the wing surfaces. But the same result can be supplied from stationary images representing the belly and back of the aircraft. Here every square centimeter on the surface is colored, the color varying with the pressure. Instead of a stationary image, a simulation system can provide a three-dimensional representation, which an engineer can manipulate on-screen to observe its surface from all possible points of view. The simulation can also provide a dynamic representation, similar to an animated drawing, displaying turbulence, pressure, temperature, and other important variables as a function of wind velocity. An engineer can even modify certain parameters of the wind description, or the shape and dimensions of the aircraft, and view the results of the changes on-screen in near real time. Almost imperceptibly we have made a transition from the relatively simple concept of digital simulation to that of *interactive graphic simulation*. The simulated phenomenon is displayed on-screen; we can manipulate variables of the model in real time and observe the resulting transformations at once. This allows us to graphically and interactively simulate complex or abstract phenomena for which no natural "image" exists: demographic dynamics, the evolution of biological species, ecosystems, wars, economic crises, the growth of a company, budgets, and so on. Here the modeling system visually and dynamically translates ordinarily invisible aspects of reality and could be equated with a very special kind of mise-en-scène.

Such simulations can be used to describe phenomena in nearly unlimited variations: predicting a set of consequences and implications for a hypothesis, obtaining a better understanding of objects or complex systems, exploring the ludic aspects of fictional universes. It is important to remember that all such simulations are based on descriptions or digital models of the simulated phenomena and are therefore only as good as those descriptions.

Places

Jeffrey Shaw's second installation piece during Artifices 1996 was called Places. In the center of a large, cylindrical room, we see a turret on which the visitor can pivot a kind of gun that projects a 120-degree image against a circular wall that serves as a screen. After familiarizing himself with the device (how to turn it left or right, move it backward or forward), the visitor begins by exploring the universe before him. This involves a complex of eleven flattened cylinders, similar in shape to the room in which the installation is located. When the visitor (virtually) penetrates one of these cylinders, a special command enables him automatically to position himself in the center and create a panorama. As it makes a complete revolution, the gun projects the panorama "contained" in the cylinder on the wall of the room. The scene reveals an industrial landscape with large reservoirs of gas and petroleum or, in another cylinder, a magnificent view of snow-capped summits and alpine forests. The visitor on the turret "turns" with the image gun so that he always faces the projected image, but behind him, 240 degrees of the circular mural screen remain blank. The visitor is put in the situation of having to "create" and "project" the explored image. However, this image has no permanence independent of his sensorimotor acts, which actualize it.

If you move straight ahead in this virtual world, you realize that it is fundamentally circular. For even though the cylinders appear to be arranged on an infinite plane, once you've gone beyond the eleventh, you find yourself again back at the first. The "curved" structure of this virtual territory, like the circular device used to actualize panoramas, illustrates the characteristics of the "new images" of cyberculture: images without edges, or frames, or limits. You are immersed in a visual universe enclosed on itself, which envelops you as you bring it into being. Behind you, there is nothing. But the moment you turn around, you create the image and reconstitute a seamless world.

The visitors around you are interested in the device. They want to hold the controls, explore the virtual world by pivoting the turret, as if they were driving an assault tank in the desert. But they quickly tire of it. "Sure it's interesting. But what's he trying to say?" They leave and others take their place. A moment earlier those same visitors, who had been waiting patiently in the room, stood between the image gun and the wall, projecting their shadow on the virtual landscape.

We often expect the arts of the virtual to hold some special fascination, to be immediately comprehensible, intuitive, devoid of culture. As if the novelty of the medium should cancel out temporal depth, the density of meaning, the patience of contemplation and interpretation. Cyberculture is not, however, the same as zapping a remote. Before we find what we are looking for on the World Wide Web, we have to learn how to navigate and familiarize ourselves with the subject. Before we can become integrated in a virtual community, we must get acquainted with its members and be recognized as one of them. Interactive works and documents generally provide no *immediate* information or emotional charge. If you don't ask questions, if you don't take the time to explore and understand, they remain sealed. It's the same with the virtual arts. No one is shocked to discover that we require an understanding of the Christian saints before we can grasp the significance of the religious frescoes of the Middle Ages, familiarity with the esoteric speculations of the Renaissance or Flemish proverbs to read a Bosch painting, or at least a minimum of mythology to grasp the subject of Rubens's canvases.

I couldn't help think of this as I listened to the disappointed comments around me. Few of the visitors seemed to have recognized the tree of the sephiroth from the Kabbala within the virtual world presented by Jeffrey Shaw. The diagram of the tree is also printed in the form of a map of the virtual world, next to the gun's controls. Indeed, the arrangement of the cylinders is identical to that of the sephiroth (the dimensions of the divine) in the diagrams of Jewish mystical tradition. Moreover, each panorama contained in the cylinders illustrates the signification of the corresponding sephira. For example, the mountain landscape corresponds to the sephira keter 'elyon, which evokes contact with the infinite and transcendence; the panorama of the great industrial reservoirs expresses the sephira malkhut, representing immanence, the reserves of energy and the benefic treasures that God has destined for the creatures of the Earth.

With this work, Jeffrey Shaw wanted to project a virtual world that was neither the representation nor the simulation of a physical, realistic three-dimensional world (even imaginary). The visitor is invited to explore a diagrammatic and symbolic space. Here the virtual world refers not to an illusion of reality but to another virtual world, one that is nontechnical and eminently real, although never "there" in the same way a physical entity is there. There is no sign of representation in the work of Jeffrey Shaw. His photographic landscapes symbolize the unrepresentable, and the respective arrangements of the cylinders enable us to read the abstract relationships among the attributes or energies of the primordial Adam. The only trace of a concrete presence in the installation is caused by the shadows of the visitors who puncture the virtual image, the inadvertent traces of the living that disturb the symbolic order. They call to mind a sentence from the Talmud: God is the shadow of man.

The Scale of Virtual Worlds

Some information systems are designed

- to *simulate* interaction between a given situation and a person
- to provide the human explorer with fine real-time control of his or her representative in the model of the simulated situation

Such systems provide the explorer of the model with the subjective sensation (although rarely the complete illusion) of personally and immediately interacting with the simulated situation.

In the previous example of simulating wind resistance on the wings of an airplane, the explorer would be able to modify the angle of view, the display of pertinent variables, the wind speed, and shape of the aircraft. The explorer, however, is not represented within the model; he or she interacts with it from the outside. But what about a flight simulator? Here the apprentice navigator sits in a control cabin that resembles an actual cockpit. The dials and screens are identical to those found in real cockpits; the control levers and switches are like those used in an actual airliner. But instead of flying an airliner, his or her actions supply data to a computer program simulating flight. Using the incoming data transmitted by the apprentice pilot and highly accurate digital models of the aircraft and its geographic position, the program calculates the position, speed, and direction a real plane would have in response to the pilot's commands. The calculations are performed at lightning speed. The simulation system projects on-screen the exterior landscape the pilot would see and displays the actual readings that would be displayed.

Virtual Reality

In the strongest sense of the term, "virtual reality" refers to a particular type of interactive simulation in which the explorer has the physical sensation of being immersed in a situation defined by a database. The effect of sensory immersion is obtained through the use of a special headpiece and data gloves. The headpiece contains two screens, located a few millimeters away from the wearer's eyes and used to provide stereoscopic vision. The images displayed on these screens are calculated in real time based on the movement of the wearer's head, in such a way that he or she can explore the digital model as if it were situated "inside" or "on the other side" of the screen. Stereo headphones complete the feeling of immersion. For example, a sound the wearer hears on the left will be heard on the right if he or she rotates 180 degrees. The data gloves are used to manipulate virtual objects. In other words, the explorer sees and feels that the image of his or her hand in the virtual world (the virtual hand) is controlled by the actual movements of his or her hand and can alter the appearance or position of virtual objects. Simple movements of the hand transform the contents of the database; this modification is immediately sent back to the explorer in a way he or she can sense. The system calculates in real time the images and sounds associated with the modification that occurred in the digital description of the situation and transmits these images and sounds to the eyepiece and headphones worn by the explorer. Various technical procedures (mechanical, magnetic, optical) are used to capture the movements of the explorer's head and hands. Tremendous processing power is needed to calculate high-definition images in real time, which explains the sketchy character of many virtual worlds at this time (1996). There has been considerable research activity in ways of improving the visual and sound quality of virtual reality systems and providing much finer tactile and proprioceptive control.

By maintaining a sensorimotor interaction with the contents of computer memory, the explorer obtains the illusion of a "reality" in which he or she is immersed—the reality described by digital memory. The explorer of virtual reality can never forget that the sensory universe in which he or she is immersed is only virtual. There are several reasons for this: images and sounds do not yet have the definition they have in cinema, there is still a slight delay between movements and their sensory impressions, equipment is heavy, and, most importantly, the explorer *knows* that he or she is interacting with a virtual reality. Like film or television, virtual reality is a *convention*, with its own codes and entrance and exit rituals. We can no more confuse virtual reality with ordinary reality than we can confuse a film or game with "true reality."

Virtuality and Information Systems

An artificial world can faithfully simulate the real world, but on a scale that is either immense or minuscule. It can allow the explorer to construct a virtual image that is very different from his or her physical appearance in everyday life. It can simulate imaginary or hypothetical physical environments, governed by laws that are unlike those governing the ordinary world. It can also simulate nonphysical spaces, symbolic or cartographic, which enable us to communicate through a universe of shared signs.

A map is not a realistic photograph but a semiotization, a useful description of a territory. By analogy, a virtual world can be compared to a map rather than to a trace or an illusion. Moreover, the territory mapped or simulated by the virtual world is not necessarily the three-dimensional physical universe. It can include abstract models of situations, universes of relations, complexes, significations, understandings, hypothetical games, or even hybrid combinations of all these "territories."

In a much weaker sense than that implied by a "realistic" sensory illusion, the concept of a virtual world doesn't necessarily imply the simulation of physical spaces or the use of heavy, expensive equipment such as stereo headpieces and data gloves.

The two distinctive—and, in this sense, much weaker—traits of the virtual world are immersion and proximity navigation. Participants are immersed in a virtual world; that is, they possess an image of themselves and their situation. Each act by the individual or group modifies the virtual world and their image in that world. In proximity navigation, the virtual world *orients* individual or group actions. In addition to traditional search and retrieval tools (indexes, hypertext links, keyword searches, etc.), cross-references, searches, and communication all occur in *proximity* to one another within a continuous space. A virtual world, even one that is not realistic, is fundamentally organized by a "tactile" and proprioceptive modality (real or transposed). The explorer of a virtual world (not necessarily realistic) must be able to control access to an immense database using mental principles and reflexes that are similar to those used to control access to his or her immediate physical environment.

A growing number of software applications and the majority of video games are based on an identical principle of real-time calculation of the interaction between a digital model of the explorer and a model of a situation, the explorer controlling the behavior and gestures of the model that represents him or her in the simulation.

Computer Virtuality

An image is said to be virtual if it originates in a digital description stored in a computer memory. Note that to be perceived, the image must appear on-screen, be printed on paper, or flash by on film: the binary code must be translated somehow. If we wanted to maintain a parallel with the philosophical meaning of virtuality, we could say that the image is virtual in the computer's memory and actual on-screen. The image is even more virtual when its digital description is not a stable deposit in computer memory but is calculated in real time by a computer program from a model and a stream of input data.

Hypertexts, hyperdocuments, simulations, and generally all software objects, such as computer programs and databases and their contents, are examples of a weakened sense of computer virtuality. This virtuality, the result of digitization, designates the process of automatically generating or calculating a large quantity of "texts," messages, visual, auditory, or tactile images, based on an initial matrix (program, model) and an ongoing interaction.

To the viewer, an animation projected in a theater or on a TV screen, even if it's produced by a computer, is the same as an animation drawn by hand. That certain special effects betray their digital origin doesn't alter the nature of our relationship to the image. It is only the production team that has really had anything to do with virtuality. In a video game, however, the player is directly confronted with the virtual character of the information. The same game cartridge (virtually) contains an infinite number of elements, or different image sequences, only some of which the player will ever actualize.

Technical manuals are supplied with industrial equipment. These manuals display on their pages text, drawings, captions, an index, the totality of information they contain. Everything is accessible to the reader. If the equipment is sufficiently complex (a fighter aircraft, space capsule, nuclear power plant, refinery, etc.), it will be impossible to provide a list of every possible fault that might occur. The manual must limit itself to providing examples of typical faults and indicating guidelines for problem resolution in other situations. In practice, only experienced technicians will be able to correct faults.

In the information technology field, however, an expert system for the same equipment may explicitly contain no more than a few hundred or few thousand rules (stored on a handful of pages). In a given situation, the user supplies the system with "facts" describing the problem. Using a rules base and these facts, the software reasons according to the

TABLE 2 The Different Meanings of the Virtual from Weakest to
        Strongest

Meanings of the Virtual	Definition	Example
The virtual as it is commonly understood	Something false, illusory, unreal, imaginary, possible	
The virtual as understood by philosophy	Exists potentially, not actually; exists without being there	The tree in the seed (compared to the actuality of a tree that has grown) A word in a language (compared to the actuality of its pronunciation)
The virtual world as understood by information technology	Universe of possibles that can be calculated from a digital model and inputs supplied by a user	Set of messages that can be delivered respectively by • software for writing, drawing, or music • hypertext systems • expert systems • interactive simulations, etc.
The virtual world as understood by information systems	The message is a space of proximity interaction in which the explorer can directly control a representative of himself	• dynamic data maps that present information based on the "viewpoint," position, or history of the explorer • networked role playing • video games • flight simulators • virtual realities, etc.
The virtual world as understood in the narrow technological sense	Illusion of sensori-motor interaction with a computer model	Use of stereo glasses, data gloves, or suits for visiting regenerated historical monuments, training for surgical operations, etc.

context and develops a precise answer (or range of answers) appropriate to the user's situation. Consequently even novices are able to correct faults. Had it been necessary to print (actualize in advance) all the situations, all the reasoning processes, and all the answers, we would have needed millions or billions of pages, far too many to manage. It is the virtual character of the expert system that makes it a more advanced instrument than a simple paper manual. Its answers (practically infinite in number) preexist only virtually. They are calculated and actualized as needed.

A virtual world in the weak sense is a universe of possibles, which can be calculated from a digital model. By interacting with the virtual world, users explore and actualize it at the same time. When these interactions are capable of enriching or modifying the model, the virtual world becomes a vector of collective intelligence and creation.

Computers and networks, then, appear as the physical infrastructure of the new informational universe of virtuality. The more they expand, the greater their power, storage capacity, and bandwidth, the greater the number of virtual worlds and the more varied they become.

CHAPTER 4

Interactivity

Beyond Pages

Characterized by finesse, delicacy, and humor, Masaki Fujihata's
Beyond Pages is one of the most beautiful illustrations of the
emerging field of "interactive art."

You enter a small space. Before you is a real table on which
is projected the image of a book. Toward the back of the room,
the image of a closed door is projected. You sit at the table and
grab a kind of electronic pencil. Using this pencil, you "touch"
the image of the book. The image of the closed book now
becomes an image of an open book, as if you had "opened" the
book. There is no actual paper book, but a succession of two
images controlled by an interactive device. Masaki Fujihata's
Beyond Pages is not a classic, stationary image. Nor is it an animation that unfolds continuously before you. It's a strange objet,
part sign (an image), part thing (you can act on it, change it,
explore it within certain limits). We're accustomed to interacting
with computer screens from our familiarity with video games,
the Internet, and CDs, but here the interactive image of the book
is located on a wooden table rather than a cathode-ray screen.

When you open this strange book, you find the word
"apple" written in English on the right-hand page and in Japanese kanji. Until this point there is nothing unusual: just written

signs on a page. But on the left-hand page, a picture of a beautiful red trompe l'oeil apple appears, whose shadow is clearly visible on the immaculate surface. It's as if the right page presented you with signs, and the left page with an object. The sensation that the apple is actually an object on a page and not simply an image is reinforced as you flip through the book. On the following page, someone has taken a bite out of the apple, and as you continue your "reading," it is gradually consumed, until you get to the end, where there is nothing on the page but a core. Each time you turn a page, you distinctly hear the sound of someone taking a bite out of the apple. The visual trompe l'oeil is accompanied by an auditory "trompe l'oreille." Yet at no time are you fooled by the illusion. You know that it's only an image and a recorded sound. It's impossible for you to take a bite out of that apple. Eating the apple appears to be a metaphor for "reading a book." Something has been consumed, something irreversible has occurred, but nothing has changed: the pages are still there, the signs as well. Unlike apples, signs aren't destroyed when we consume or enjoy them.

This oscillation between sign and thing, the sign that sounds, acts, interacts, and appears to disappear like a thing, the thing that is as impalpable and indestructible as a sign, continues until you finish "reading" the book. The stones you move with your pencil make a scratching sound against the image of the paper. When you activate the image of a handle on the page, the door on the back wall opens, and an adorable young girl emerges, naked and laughing. You make her reappear several times.

Unlike the dried plant leaves found in herbariums, the branch of green leaves that trembles in *Beyond Pages* moves with the wind and swells with sap. The dried flower or leaf is there before us, dead but quite real, between its pages. Yet *Beyond Pages* leads us somewhere beyond the page, where "living" images of living things seem to rise from its pages.

At the end of the book, the flowering signs begin to speak.

Your doodles are miraculously transformed into perfectly drawn Japanese writing and clearly spoken by the "book." The book speaks too. It has a voice that enables it to read itself, and you are invited to help write it.

Beyond Pages is partly based on the Möbius strip, the continuous but unnoticed transition from one order of reality to another: from the sign to the thing, from the thing to the sign, from the image to the character, from the character to the image, from reading to writing, from writing to reading. It's the image of a book (and thus a sign twice over) between whose pages things are found ... which are, in the end, only signs, but active, living signs that answer back. The virtual is not an illusion of reality, as it is so often described, because the participant is always aware that it is a game, an artifice. Rather, it is the ludic or emotional truth of an illusion that is experienced as such.

Interactivity as Problem

Because interactivity is so often misrepresented, as if everyone knew exactly what it was about, in this short chapter I'd like to offer instead a *problematic* approach to the concept.

The term "interactivity" generally refers to the active participation of the beneficiary of an information transaction. Indeed, we would be hard-pressed to demonstrate that an information receiver—unless dead—is ever passive. Even seated in front of a television without a remote control, the recipient decodes, interprets, participates, and mobilizes her nervous system in a hundred different ways, and always somewhat differently than the person sitting next to her. Satellites and cable provide access to hundreds of different channels. Linked to a VCR, this enables the viewer to put together a video library and define a televisual system that is obviously more "interactive" than a single TV channel alone. The possibility of material reappropriation and recombination of the message by a receiver is a significant factor in evaluating the degree of interactivity of a device. The same is true of other media: can we add nodes and links to a hyperdocument? Can we connect this hyperdocument to others? With television, digitization could further enhance the opportunities for reappropriating and personalizing the message by shifting editorial functions to the user: choice of camera, use of zoom, shifting between image and commentary, selection of commentators.

Does this mean that in an interactive environment, the communications channel operates in both directions? In this case, the paragon of an interactive medium would unquestionably be the telephone. It supplies dialogue, reciprocity, and real communication, whereas television, even digital TV, which can be navigated and recorded, provides only entertainment. Even a classic video game is more interactive than television, although the game doesn't really provide reciprocity or communication with another person. Far from an uninterruptible stream of images rolling across the screen, the video game reacts to the player's actions, and the player reacts in turn to the images before him: interaction. The TV viewer zaps and selects; the player acts. Yet the ability to interrupt an information sequence and accurately reorient the information flow in real time is not just a characteristic of video games and digital hyperdocuments; it's also a characteristic of the telephone. Only in one case we're speaking to another person, and in the other with an information matrix, a model capable of generating an almost infinite number of "parties" or different paths (all of which are coherent). In this case, interactivity is a lot like virtuality.

Let's look at the differences between games and the telephone from another angle. To compare apples with apples, I'll assume that a networked game enables two adversaries to play against each other. This arrangement maximizes the similarity between the game and the telephone. In the video game, each player, equipped with a joystick, data gloves, or some other device, modifies *his own image in the game space.* The character tries to dodge a shell, move toward a goal, explore a passage, win or lose weapons, powers, "lives." It is this modified image of the reactualized character that then modifies, in logical time, the game space itself. The player isn't fully involved in the game unless he projects himself into the character that represents him and, therefore, into the field of danger, force, and opportunity in which he exists, within the shared virtual world. With each "move," the player transmits to his partner a different image of himself and a different image of their shared world, images the partner receives directly (or can discover through exploration) and is immediately affected by. The message is a doubled image incorporating the situation and the player.

During a telephone call, however, speaker A transmits to speaker B a message that is supposed to help B construct, through inference, an image of A and the situation shared by A and B. B does the same with respect to A. The information transmitted during the ebb and flow of communication is much more limited than it is in a virtual reality game. The equivalent of the game space, that is, the context or situation comprising the respective position and identity of the partners, is not shared by A and B as an explicit representation, a complete and explorable image. The reason is that the context is here unlimited a priori, whereas it is circumscribed in the game, something that is also a function of the difference in the nature of the communications devices themselves. With the telephone, the reactualized image of the situation must constantly be reconstructed by the speakers, separately and individually. The videophone doesn't alter this in the slightest, since the true context, the universe of signification, the pragmatic situation (resources, the field of forces, threats, and opportunities, everything that can affect the plans, identities, or survival of the participants), is not shared to any greater extent simply by adding an image of the physical appearance of the person and the immediate physical environment. Systems that provide shared, remote access to documents, sources of information, or work spaces begin to approximate communication in a virtual world, until we reach systems that incorporate one or more active images of the person (filtering software agents, information trackers, personalized profiles, etc.).

Communication through a virtual world is in one sense more interactive than telephone communication because it implies within the message the image of the person and the situation—nearly always the key elements of communication. Yet in another sense, the telephone is more interactive, because it puts us in touch with the *body* of the speaker.

Not an image of her body but her voice, an essential dimension of her physical manifestation. The voice of the person I'm speaking with is truly present wherever I happen to be talking to her. I'm referring not to an image of her voice but to her voice itself. Through this physical contact, an entire affective dimension is "interactively" transported during a telephone call. The telephone is the first telepresence medium. Today a number of R&D projects are attempting to extend and popularize telepresence to other physical dimensions: remote manipulation, three-dimensional images of the body, virtual reality, augmented reality environments for videoconferences, and others.

We see from this brief overview that the degree of interactivity of a medium or communications system can be measured using several different criteria:

- the ability to appropriate and *personalize* the received message, regardless of the nature of the message
- *reciprocity* of the communication (a one-to-one or many-to-many communications system)
- *virtuality*, here understood in terms of the processing of the message in real time based on a model and input data (see table 2)
- the *incorporation* of the image of the participants in the message (see table 2)
- *telepresence*

In table 3, for example, only two axes intersect among the many we could have used to highlight the concept of interactivity.

Through the virtualization of information, progress in interface design, and increased computing power and bandwidth, hybrid and mutant media proliferate. Each individual communications system must be subject to precise analysis, which is itself based on an updated theory of communication or, at the very least, a more detailed map of the modes of communication. The preparation of such a map has become increasingly urgent as political, cultural, aesthetic, economic, social, educational, and even epistemological issues become ever more dependent on the ways in which communication systems are configured. Interactivity has more to do with finding the solution to a problem, the need to develop new ways to observe, design, and evaluate methods of communication, than it does with identifying a simple, unique characteristic that can be assigned to a given system.

TABLE 3 The Different Types of Interactivity

Communication system	Linear message, not modified in real time	Interruption and reorientation of the information stream in real time	Involvement of the participant in the message
One-way distribution	press radio television cinema	• multimodal database • static hyperdocuments • simulation without immersion or the ability to modify the model	• single-participant video games • simulation with immersion (flight simulators) but without possible modification of the model
Dialogue, reciprocity	mail correspondence between two people	• telephone • videophone	dialogues that take place through virtual worlds, cybersex
Multilogue	• correspondence networks • publication systems in a research community • e-mail • electronic conferences	• multiparticipant teleconference or video conference • open hyperdocuments accessible on-line, written and read by a community • Simulation (with the ability to act on the model) as a medium for community debate	• multiuser role playing in cyberspace • multiparticipant video games in "virtual reality" • communication through virtual worlds, continuous negotiation of participants with their image and the image of their shared situation

CHAPTER 5

Cyberspace, or
The Virtualization of Communication

Navigating the World Wide Web

This book is not a practical guide to navigating the World Wide Web but an essay on the cultural implications of the development of cyberspace. Readers who are interested in the practical aspects of the Internet but have no personal experience can refer to any of a number of practical handbooks on the subject. Many magazines provide lists of the best sites on the Web, organized by topic and updated periodically. But as soon as we enter the universe of the Web, we discover that it is not only an immense "territory" that is expanding at an ever accelerating rate but one that provides users with a vast array of maps, filters, and tools for finding their way. The best Web guide is the Web itself. Its exploration requires patience, and we must be willing to risk getting lost, to waste time in familiarizing ourselves with this strange land. We may need to give in to the Web's playful side if we want to discover the sites that are most related to our professional or personal interests and therefore capable of optimally satisfying our personal journey in cyberspace.

There are two primary, but opposite, approaches to navigating the Web, actual searches being generally a mixture of the two. The first might be referred to as *hunting*. This method is applicable when we are looking for specific information and want it as quickly as possible. The second might be referred to as *gathering*. Vaguely interested in a subject, but ready to veer off at any moment onto a different path, not knowing exactly what we're looking for but nearly always finding something in the end, we wander from site to site, link to link, collecting what we need along the way.

Since we can find nearly everything and anything on the Internet (or if not actually everything, then certainly references to everything), any specific example would be incomplete, and no one can provide an idea of the infinite number of navigations possible. One of my friends, who recently bought an old harmonium and wanted to have it repaired, told me that he found two sites on the Net that gave detailed explanations for repairing the instrument. He also located a newsgroup that provided answers to his last nagging questions. Except for the parts, the harmonium was repaired without cost.

Because every navigation is unique, I can't provide any significant examples of Web surfing. Instead I'll describe my last two sessions on the Internet; one was devoted to hunting, the other to gathering.

Hunting

One evening my companion said to me, "I haven't heard from Olaf Mansis (not his real name) in three years. You claim we can find anything on the Internet, so how can I get in touch with my friend?"

My interest was piqued. I asked her if she thought he had an e-mail address, since there are specialized search engines designed to locate people on the Internet. She told me she was almost sure he didn't, since he was one of those people who hated new technology. We would have to try a different approach, then. After making sure my modem was connected, I clicked an icon on my desktop and initiated the procedure that would connect me to my Internet service provider. The modem squawked, indicating that it had made contact. Once on-line, I opened a specialized software application, known as a browser, designed for navigating the World Wide Web. In my browser, I checked my list of favorite sites (sites I use most frequently). The list included Alta Vista, which is one of the most popular search engines on the

Web. I clicked to connect to their Web site, and after a few seconds, the Alta Vista home page was on my screen—along with a few ads, which I ignored. In the search box I entered the words "Olaf AND Mansis." This would find all documents containing both character strings.

A few seconds after starting the search, I received an answer: there were fourteen sites that satisfied my search criteria. By clicking each item in the list, I was able to access the corresponding site. I took my time and looked at each of the sites, one after the other. Some of them referred to highly specialized medical conferences in Swedish, German, or English, at which either a Mr. Mansis or an Olaf had spoken. This was certainly a dead end, since the Olaf Mansis we were looking for is a painter. We looked at site after site, but none of them had anything to do with our Olaf.

Finally, on the twelfth site on the list, which was for a Canadian auctioneer, we found a list of works sold at auction, including a painting signed by our friend. Excited, we sent a message to the e-mail address provided for the auctioneer, telling him of our problem and asking if he knew where we could get in touch with Olaf. On the last site, a Dutch university, appeared a list of students, along with their e-mail addresses. There were several Olafs, along with a Margaret Mansis. Since Olaf Mansis is Dutch, we were hoping Margaret was a relative of his, even a remote one, who might be able to give us some idea of where he was. We sent her an e-mail message as well.

Two days later, we received a response from the auctioneer, who was unable to help us. Although he was selling the painting, he didn't know how to get in touch with the artist. A week later, however, we received a message from Margaret Mansis, who was indeed related to the painter (a grandniece, in fact), indicating that she would try to help us—providing we gave her some information about ourselves and the reasons for our search. After an exchange of e-mail, we received the information about Olaf from our correspondent, who had obtained it during a family dinner.

We see from this example that even if we can't obtain the information we're looking for directly from the Internet, we may be able to locate people or organizations who can help us find it. It's also worth noting that our initial search took fifteen minutes and was conducted at home for the price of a local phone call. Imagine the time and effort required to do this manually, including an examination of all the telephone books on the planet. Recall that the purpose of the operation was to help two friends who had lost sight of each other get back in touch. The virtual doesn't replace the real; it simply increases the opportunities to actualize it.

Gathering

From time to time, I visit the Virgin Paris Web site to read reviews (updated every month) of the latest records. Music, books, and videos are classified by style, and I generally look at the pages for "gaia" and "techno." In this month's gaia section, I found a new album by Gavin Bryars, *Farewell to Philosophy*. Near the description of the album, there is usually an icon allowing you to download an audio sample from the record, along with one or more links to sites for musicians. I followed the link to the Gavin Bryars site, and after a few seconds, I had access to a biography and complete discography for the artist. I learned that he had made far more albums than I was aware of. I wrote down the titles that looked interesting and returned to the Virgin record site.

This ability to learn more about a topic that is merely hinted at, through a direct link to a more specialized site (which can be located physically anywhere in the world; Gavin Bryars's site was in England), is a unique feature of the Web and one of its most remarkable advantages.

From the Virgin home page, I followed a link to a debate about current events, specifically a court sentence against the group NTM for having insulted the police. To spur the discussion, the site's publishers had provided access to a number of French songs from all periods that had (verbally) abused the forces of law and order. Half-amused, half-shocked, I read the contributions concerning NTM's sentencing. The debate was extremely lively, and in spite of a handful of peremptory and poorly argued statements, I found that the comments and ideas were extremely well reasoned on both sides of the argument. Back on the home page, I followed a link to an imaginary dada museum. It consisted mostly of links to sites on dadaism, surrealism, and nonsense verse, along with photographs of works by Max Ernst, Marcel Duchamp, Man Ray, and others. There was also information about dada-inspired musicians such as Frank Zappa, with links to other sites detailing the life and work of these musicians. The imaginary dada museum itself contained links to a number of other sites, including the following:

- the Web Museum (one of the oldest and most remarkable virtual museums)
- the Surrealism Server
- the From Dada to Wave site
- several sites about Alfred Jarry and Ubu
- the Fluxus site
- a site devoted to the Situationist International

I decided to follow some of the links and was surprised to discover on the Situationist International site a review of an article by Bruno Latour that had appeared in *Libération* on the suicide of Guy Debord (the author of *The Society of the Spectacle* and one of the principal founders of SI).

On the Fluxus site I found photographs of installations and reports of performances, one more bizarre than the next. In particular, there was a reproduction of a *Vagina Painting* that left me dumbstruck. I was happy to finally come across an installation by Naim June Paik, whom I had heard about but with whose work

I was only vaguely familiar. There was a Buddha (Naim June Paik is Korean) meditating in front of a video monitor. On top of the screen was a camera, which filmed the Buddha … and displayed the image of the Enlightened One live on-screen. I was immediately struck by the installation's ability to provide a much better practical illustration of meditation than the traditional statue. The statue is a stationary image, a massive, solid block, nearly timeless. Yet meditation is a form of attention that is constantly renewed in the present. It is well depicted by the concept of a real-time loop and an ongoing process—although nothing seems to happen—in Paik's installation.

The Fluxus site also contains links to other art sites, primarily the Media Filter site, which organizes a forum and maintains an archive on police misconduct, supplied by volunteer contributors around the world.

Abandoning the other links, I decided to go straight to the Rhizome site from my list of favorites. Rhizome, which has correspondents pretty much all over the world, sends out a free weekly message to subscribers (like myself) containing information and commentary about contemporary art. The site presents articles and links to other sites, which provide information about the work discussed in the articles. This may be "classical" artwork, like the portraits on aluminum plate made from fragments of genetic code, found on the artist's site. But there are also references to works whose natural environment is the Internet. The virtual world constructed by Julia Sher on Ada Web plays on the masochism and paranoia of Web surfers. Visitors become lost in an interactive universe of terrifying clinics, surveillance, and chat groups on the pleasure of being dominated. Because the contributions are so well written, one wonders whether the participants (whose sex is unknown) are real or fictional.

I completed my navigation at the site of a Brazilian friend

living in the United States, who had told me by e-mail that one of his new works, *Rara Avis*, was now on the Internet. This was a real-time video image (refreshed every sixty seconds) of a birdcage containing a rare species of parrot. The image changes from one minute to the next. This type of experimental work, which is becoming increasingly common on the Internet, prefigures a completely different relationship to the video image than we now have through television. We will soon have "direct" access to an increasing number of places on the globe whose image is available on the Web. I've seen real-time pictures of NASA labs, an intersection somewhere in Washington, a remote part of Antarctica, the Earth seen from a satellite.

Obviously I could have provided other examples, including one during which I was able to download the writings of Nagarjuna (the great Buddhist philosopher of the "middle way") and another in which I helped compile a lexicon of cyberphilosophy. But I had to make a choice and preferred to use a conventional example, my most recent session, chosen more or less at random.

My Web session took a little more than an hour, and in my opinion this was time well spent, far more worthwhile than the time it would have taken to read a couple of print reviews. "Gathering" on the Internet can only be compared with wandering around an immense illustrated library that is easy to access, provides real-time delivery and interactivity, and is participatory, impertinent, and playful. This media library is populated, global in scope, and constantly growing. It contains books, records, TV and radio programs, magazines, journals, propaganda, CVs, video games, discussion and meeting areas, and markets, all of which are interconnected, living, fluid. Far from being uniform, the Internet is becoming increasingly global, with more languages, cultures, and variety each year. It is up to us to continue to supply it with diversity and exercise our curiosity so that the pearls of wisdom and pleasure it harbors are not buried at the bottom of the informational ocean.

What Is Cyberspace?

The word "cyberspace" was coined in 1984 by William Gibson in his science fiction novel *Neuromancer*. The term refers to the universe of digital networks, described as a battlefield among multinationals, the site of global conflict, the new economic and cultural frontier. In *Neuromancer* the exploration of cyberspace introduces us to hidden fortresses of information, protected by ramparts of software, islands bathed by oceans of data, which interact and mutate at high speed around the planet. Some of the characters in the book are able to "physically" enter this data space. Gibson's cyberspace makes us aware of the geographic movement of information, which is normally invisible. The term was immediately adopted by the users and designers of digital networks. Today there are any number of literary, musical, artistic, and even political currents that claim to be part of cyberculture.

I define cyberspace as the *communications space made accessible through the global interconnection of computers and computer memories*. This definition includes all electronic communications systems (including wireless networks and conventional telephone systems) to the extent that they convey information from digital sources or sources intended for digitization.[1] I insist on the use of digital encoding, because it conditions the plastic, fluid, calculable, real-time, hypertextual, interactive, and, yes, virtual nature of information, which is in my opinion the distinctive characteristic of cyberspace. This new medium is designed to serve as an interface to existing systems for creating and recording information, communication, and simulation. The general digitization of information will most likely turn cyberspace into the principal communications channel and primary medium of humanity's memory by the year 2000.

In the following sections, I want to briefly discuss the principal modes of interaction and communication in cyberspace. It should be obvious that everything that has been provided by traditional television, radio, and telephony can also be implemented through the use of digital equipment; there is no need to belabor this point. I will focus instead on the major innovations of digital technology in comparison with earlier communications systems.

Remote Access and File Transfer

One of cyberspace's principal functions is to provide remote access to various computer resources. For example, assuming I have the proper authorization, I can use my personal computer to connect to a larger machine thousands of miles away and perform operations (scientific calculations, simulations, image generation, etc.) in a matter of minutes that my small PC would take days to execute. This means that cyberspace can supply computational power in real time, a bit like the way large utility companies distribute energy. From a strictly technical point of view, we no longer need a powerful computer of our own; we simply require access to computational power accessible somewhere in cyberspace.

With an appropriate terminal (PC, enhanced TV, cellular phone, personal digital assistant, etc.), I can also access the contents of a database or, in general, the data stored in a remote computer. Providing I have the needed software interfaces and sufficient bandwidth, it's as if I were reading data from my own computer's hard drive. If the cost of the connection is low, there is no need for the information to be stored locally. As soon as public information finds its way to cyberspace, it is virtually and immediately available to me, independent of the spatial coordinates of its physical medium. Not only can I read a text, navigate a hypertext, watch a sequence of images, display a video, interact with a simulation, or listen to recorded music, but I can also supply this remote archive with text and graphics. Dispersed communities can communicate with one another by sharing a remote memory store, which each member can read from and write to, regardless of geographic location.

Cyberspace also makes it possible to transfer files or download information. Transferring a file consists in copying a packet of information from one digital storage area to another, generally from a remote storage site to my personal computer or the site where I'm physically working. The information transferred from CERN's computer in Geneva to the PC of a physics student in Melbourne doesn't disappear from CERN's computer. The file could be the latest update of a database containing the most recent results of the high-energy physics lab, or a bank of photographs of elementary particle collisions in a bubble chamber, or the text of a scientific article, or an educational video about an installation, or an interactive model illustrating the theory of hyperstrings, or an expert system for diagnosing cyclotron faults. The student in Melbourne can copy these files only if they have been classified by the CERN administrators as public information, which can be downloaded by anyone. Otherwise a password will be required, or the information will be supplied on a paid subscription basis.

Obviously, computer software is one kind of information that can be downloaded remotely. In this case, downloading provides a method for rapidly distributing the tools (software) for improving the operation of cyberspace, using cyberspace itself as the distribution channel. A large number of software programs for optimizing communications among computers and searching for information in cyberspace have been distributed in this way.

Electronic Mail

Messaging is one of the most important and frequently used features of cyberspace. Anyone connected to a computer network can have access to an electronic mailbox identified by a special address and can receive and send messages to anyone with an electronic mail address that is accessible from the sender's network. These messages are principally text based but will become increasingly multimodal in the future.

To fully understand the value of e-mail, we need to compare it to conventional mail and faxes. Messages received in an electronic mailbox arrive in digital form. This means they can easily be deleted, modified, and stored in digital format by the receiver, without the need for paper as a medium. Similarly, we no longer need to print a text to send it to an addressee; we can send it directly in its digital form. This feature is even more useful today, when so many messages originate on a computer.

Wherever the ability to connect through radio waves or over phone lines exists, even indirectly, I can send and receive messages with a computer (which today means nearly anywhere). With electronic mail, I can send the same message to a list of correspondents by simply indicating the list as the recipient. It is no longer necessary to make multiple photocopies of a document or dial one phone number after another to send a message by fax. If every member of a group has a list of addresses for other members, group-to-group communication becomes possible. Every individual can send messages to the entire group knowing that the other members of the group have received the same messages as he.

Nuclear Tests

One evening I was reading my mail, as I do every evening. I opened a message from the organizers of an important international conference on virtual art where I was supposed to appear. I was told, in English, that a mailing list would be set up so that we could begin our discussion before the actual physical meeting took place. To join, all I had to do was send a message saying "I subscribe" to the e-mail address provided. Interested, I sent the message. The following evening, in addition to my regular mail, I discovered several messages from the digital arts mailing list.

A professor from an art school in Minneapolis explained his colleagues' failure to comprehend his work on multimedia. A Dutch artist talked of his installations, where he records the sound of the sea along the coast … and the enormous artificial seashells that reflect this sound from places he has selected underground. A student in Detroit expressed fear that the multimedia industry will—for commercial reasons—standardize visual, sound, and tactile interfaces, which artists would like to leave open so that they can explore the possible alternatives.

On the following day, my mailbox contained responses to the previous day's messages. Some respondents added to the concerns of the initial writers; others contradicted them. Many artists regretted not having been invited to display their work at the conference, even though they had submitted a proposal. They took advantage of the mailing list to inform the community that there was a Web address where they could obtain a description or example of the artists' work. On the following day, one of the organizers of the conference informed the list that the budget was limited, that as many as eighty artists from around the world would be able to display their work, and that this was already a considerable number.

As the days passed, a few topics seemed to coalesce: institutional and educational questions, aesthetic problems, information about software. The majority of the messages were labeled as responses to a previous message, which were themselves responses to even earlier messages, and so on. It was relatively easy to re-create the thread of a relatively individual conversation. In time, some exchanges on a given topic included twenty, thirty, or more responses. For others there were no more than five or six responses, and the conversation died out on its own.

It is customary on the Internet to include in your message the message to which you are responding, so that an e-mail begins to look like a kind of commentary on the previous message. Several layers of text (sometimes four or five) can be found inside a message, each quoted section becoming a kind of "envelope" for the previous message. E-mail software promotes this feature by automatically reproducing (using a special symbol at the beginning of each quoted line) the message you're responding to in your reply. Some mailing list subscribers complained about the excesses of this feature, which, like a snowball rolling down a hill, artificially increased the size of messages in their mailbox.

The messages came from all over the world, although the majority were from North America and Europe. As is often the case on mailing lists, even though there may be 250 subscribers (all of whom receive messages), only a handful of individuals actively participate in the conversation. Little by little, the list subscribers discovered the individual style of these natural leaders—undoubtedly a reflection of their character. Some of these people were spontaneous and emotive and wrote a haphazard, almost phonetic English. Others responded point by point, with almost maniacal attention, to the comments of other participants or wrote what amounted to small treatises several chapters long, which were carefully composed. When emotions flared, the moderators (whom I imagined to be "older") stepped in and tried to calm things down.

From time to time, as the muffled sound of a polluted Paris beat against the windows of my apartment and my tired eyes could barely make out the characters on-screen, a subscriber would wander from the topic of discussion to talk about the weather in Oslo or a recent retreat in the mountains of Colorado. There, without a computer or access to the Net, stretched out on the flower-covered slope, he had breathed in the cool wind from the mountaintops, which carried with it the resinous scent of pine, and lost himself in the pure depths of the blue sky above.

Our routine correspondence about the conference was interrupted by a message from an Australian musician, a fellow by the name of Wesson (not his real name), who violently protested the nuclear tests being conducted by France in the Pacific. His message triggered a number of responses during the next few days. Some were sympathetic. Others reminded Wesson that the digital arts mailing list hadn't been set up to talk about nuclear testing and that there were a number of lists on the Internet where he could speak with others interested in the subject. Some participants responded that no artist could exclude any topic of discussion a priori; artists had always been involved in affairs of state, which now extended around the globe. The discussion grew bitter. Some subscribers threatened to unsubscribe from the list if the flood of messages on nuclear testing didn't stop. Wesson, increasingly excited, posted a message in which he acknowledged that he had begun to learn French but now regretted his interest in the language. This time no one supported Wesson, and he was flamed mercilessly. A flame is a written attack, a kind of verbal assault, sent by e-mail, and they generally come in bunches. Wesson was flamed by the French, Belgians, Swiss, and Quebecois in the language of Molière.

A German, Englishman, and Dane also answered in French, out of solidarity with a minority language that had been insulted. Some American academics attempted to reason with Wesson while reproaching him for having ignored the ethics of the Net. Like many of the other subscribers, who had been content merely to read the messages that were posted, I overcame my sense of reserve and addressed Wesson (in English). I explained to him that he was confusing at least two separate issues: that of a language and a people, and that of a people and its government. Wesson, who claimed to be a pacifist, should realize that it was this type of egregious error, the identification of human beings with national, ethnic, linguistic, or religious categories, that made war possible.

Wesson then made what might be described as a public confession. He regretted his message about French and asked us to accept his apologies. When he had written his unfortunate e-mail, he had been alone in front of his computer screen. It was as if he were thinking out loud, without realizing that there were people on the other side of the network—living, animate, sentient individuals, who were as susceptible to being wounded by language as he was. And among them were individuals whom the TV and newspapers had subjected en masse to the condemnation of the Australians. His anger had been brought to a white heat by the anti-French criticism of the media around him. But the network had given him a sense of global awareness that was much more concrete than he imagined, for he was in direct contact with other individuals who expressed their own emotions and thoughts. Following Wesson's message to the list, I was surprised to find a private message from Wesson in my mailbox, which other members of the list obviously could not read. He told me he had been touched by the sincerity and clarity of my response and that he would like to meet me. We then exchanged a few personal messages, which concluded with a mutual promise to get together during the conference.

> The summer passed. One morning in September, in the press room for the international symposium, a bearded young man with a smile on his face approached me.
>
> Mister Lévy?
>
> Yes.
>
> I'm Paul Wesson.

Newsgroups and Mailing Lists

More complex than e-mail, a newsgroup is a sophisticated software application that enables groups of people to discuss specific topics together. Messages are generally classified by topics and subtopics. Some topics are closed when they are empty, and others opened whenever the members of the group feel the need. In an electronic newsgroup, messages are addressed not to individuals but to a topic. This doesn't prevent individuals from replying to one another, since the messages are signed. And individuals who come into contact through a newsgroup can generally communicate through conventional e-mail as well, person to person.

Chat software enables direct communication among all the people connected to a group *at the same time*. Messages exchanged in this type of format are generally not saved. Individuals share a kind of ephemeral virtual communications space, in which new forms of writing and interaction are created.

Some message systems and newsgroups operate only on specialized networks run by large corporations or commercial services. However, the trend has been to set up gateways between these local systems and the larger system for interconnecting networks known as the Internet.

Gradually everyone with a mailbox on a computer network—soon just about anyone with a computer or PDA—will be able to receive a multimodal message from any point of entry to cyberspace, just as the telecommunications system can connect a telephone to any other on the network.

Internetworks like the Internet provide access to a large number of electronic newsgroups. By giving visibility to these discussion groups, which are continuously being formed and disbanded, cyberspace becomes a means of getting in touch with people on the basis not of their name or geographic location but of their interests. It's as if the people who participated in newsgroups had acquired an address within the moving space of discussion topics and objects of knowledge.

From Newsgroups to Groupware

When search and indexing systems are integrated and all contributions saved, newsgroups and mailing lists function a bit like group memory. We then obtain a living database, continuously fed by groups of individuals in direct contact and interested in the same subjects. At some point, the distinction between an on-line hyperdocument, which each member of the community can read from and write to, and a sophisticated electronic newsgroup becomes murky. Locked in a CD-ROM, a hyperdocument, even if it retains some of the interactive features that characterize digital communication, offers less plasticity, dynamism, and responsiveness to the changing context than a hyperdocument that is enriched and restructured in real time by a community of authors and readers on a network. Buzzing, blooming, bifurcating, a dynamic rhizome that expresses a plurality of experiences in its construction, harboring the multiple and multiply interpreted memory of a collectivity, inviting navigation among nonhierarchical strands of meaning, the hypertext only deploys the full range of its qualities when immersed in cyberspace.

Some group-learning programs are designed for sharing computer resources and use the means of communication typical of cyberspace. These are known as computer-supported cooperative learning (CSCL) programs. They are used for group discussion, knowledge sharing, and the exchange of experience among individuals, access to on-line tutors, databases, hyperdocuments, and simulations. In more advanced systems, hyperdocuments are restructured and enriched with the experiences of other learners.

New forms of organizing work have also appeared, which fully exploit the resources of shared hyperdocuments, newsgroups, remote access, and file transfer. The field of computer-supported cooperative work (CSCW) is expanding rapidly. If it is well designed, a cooperative work system on a network can *also* serve as a system for cooperative learning. Software and computer systems for cooperative work efforts are known as *groupware*. Internet tools, such as e-mail, newsgroups, mailing lists, and Web browsers, are increasingly being used on *intranets*, which are designed to facilitate communication within the enterprise. Intranets, which are becoming increasingly standardized, provide the tools for correspondence, collaboration, and sharing data and documents that are compatible with the Internet. Today the most diverse types of transaction between information systems and organizations are becoming transparent on intranets.

Cyberspace enables us to combine several modes of communication. In order of increasing complexity, it has given us e-mail, newsgroups, shared hyperdocuments, sophisticated cooperative learning or work systems, and, finally, multiparticipant virtual worlds.

The Electronic Bulletin Board:
An Example of a Virtual Community

Jean-Michel Billaud is head of the business intelligence and marketing group at Compagnie Bancaire. For several years, and long before it was popular with the press, he has understood that the development of on-line digital communication would result in profound economic change as well as a reorganization among the leading financial organizations. What would happen to commerce when a large number of our financial transactions take place in cyberspace? What would happen to national currencies when cybercash becomes commonplace? Will bankers have to reinvent their traditional role when different forms of on-line credit and exchange—still in the process of experimentation— are well established? How will they influence the changes that are now under way? Who will be capable of acting and how? These are the kinds of questions Jean-Michel Billaud began to ask within the community of French bankers and credit organizations through a limited-subscription magazine known as *L'Atelier*

de la Compagnie bancaire. Combining experimentation with
information and reflection, this visionary promoted the use of a
bulletin board in his company's department of business intelligence, encouraging other department managers to join. He
wanted to accustom corporate managers to receiving targeted,
filtered information by electronic means, to initiate them in the
culture of nonhierarchical exchange and collective decision
making using digital media. The bulletin board was soon made
accessible to people outside the bank, people who were interested
in the issues raised but also capable of supplying information
based on their own experience. Naturally, some information and
certain discussion groups were restricted to employees of the
bank. But the public section quickly grew to amazing size. The
Atelier's BBS, which has several hundred subscribers, is currently
one of the largest, if not the largest, French virtual community.

If you are, as I am, a subscriber to this BBS, you have access
to four services: e-mail (with a gateway to the Internet), news,
forums, and access to documents and software for downloading.
I'd like to provide more detail about the type of information
provided and the bank's electronic forums.

News and Information. Documentalists and information specialists who work in Compagnie Bancaire's business intelligence
group maintain well-documented files, which contain information from various Internet newsgroups, Web sites, semiprivate
e-mail, and specialized organizations. This information is
generally about information technology and telecommunications
markets: market trends, statistics, mergers and acquisitions,
investment decisions, and financial innovation on the Internet.
You can follow the day-to-day progress of the Java programming
language, used for interactive applications on the Web, or the
use of network computers, specialized hardware designed to run
on a network and intended for the mass market (and far less
expensive than a conventional PC). Some files contain bibliographies and documentation rather than raw data. Back issues of

L'Atelier, along with forthcoming articles, are also available on-line.

Providing a summary and selection of current news, the BBS's daily press summary is highly regarded by subscribers. Information is culled from specialized newsletters as well as the major dailies. For example, the headings in the press summary for 5 February 1997 included the following topics: on-line services, electronic funds transfer, mobile telephony, telecommunications, information technology, electronics, audiovisual, digital television.

Forums. As in any virtual community, the forums on the *Atelier*'s BBS are places for the exchange of information and discussion. Some forums are open to anyone; others, highly technical in nature, are frequented only by specialists. Internet newsgroups and forums from other BBSs are also available. The following is a list of some of the forums available on the bank's BBS: electronic commerce, secure payment systems, the stock market, the information superhighway, legal questions concerning cyberculture, virtual reality, bulletin boards, democracy and new information technologies, digital cities, CD-ROMs. The BBS also hosts the newsgroup for the Club de l'Arche, whose mission is to sensitize political and economic leaders to emerging issues of cyberculture. One of the goals of the *Atelier*'s BBS is to attract an increasing number of participants, promoters, and decision makers in related fields. Using the bank's BBS, a pressure group composed of French financial managers in favor of liberalizing the use of cryptography was organized. Some reserved areas, used for financial negotiations and transactions, are accessible only to individuals with a password.

I'd like to conclude this section by pointing out that reading information on-line and participating in virtual discussions does not replace flesh-and-blood contacts. On the contrary. In fact, the magazine organizes physical meetings several times a week in Paris, at which participants in the cybereconomy can mingle. New products are presented, new initiatives related to *L'Atelier*'s favorite topics are introduced, and open debates take place, during which the utopians of electronic democracy and collective intelligence can rub shoulders with marketing managers (sometimes the same individual). And all can raise a glass together with Jean-Michel Billaud.

Communication in the Shared Virtual World

As noted earlier, interaction with a virtual reality in the strongest sense reflects our ability to explore or alter the contents of a database through our movements (head, hands, body, etc.) and immediately perceive, through the senses (images, sounds, tactile and proprioceptive sensations), the new state of the database we have modified. This implies that we can maintain a sensorimotor relation with the contents of a computer memory. However, virtual realities are increasingly being used as a communications medium. Several geographically dispersed individuals can populate a database at the same time through their gestures and receive sensory information back from the database. When one of them modifies the contents of the shared digital memory, the others immediately perceive the new state of the shared environment. Because the position and virtual image of each participant is recorded in the database, every time one of the participants moves or modifies the description of her image, the others perceive her movement. This type of communications system can be used for games, learning, work, simulations of urban and combat environments. Shared virtual realities, capable of putting thousands, if not millions, of people in contact with one another, should be considered many-to-many communications systems, typical of cyberculture.

We can extend the concept of communication through a shared virtual world to systems other than those that simulate interaction within a "realistic" three-dimensional physical universe whose visual aspect is calculated according to the laws of perspective. In other words, we can communicate through virtual worlds even in a weaker sense than that implied by the use of immersive simulations.

It is not necessary that a given communications system calculate images and sound for it to be considered a virtual world. For example, certain role-playing games involving thousands of Internet participants are authentic virtual worlds although they are not composed of texts: they have rules of operation and the capability of autonomous activity. Every player helps construct the universe in which the character he embodies participates. Moving within this fictional universe, the players are more or less near one another and only interact if they are in the same virtual location. This is an excellent example of communication through the cooperative construction of a world, and obviously a kind of many-to-many system.

Navigation

Even those who know nothing about programming can use e-mail or read newsgroups or view a remote hyperdocument on a network. In general, all that is required is the ability to click the right buttons or make the correct choice from a menu. At worst we may have to type a few commands, which are rapidly committed to memory. For many years, however, moving from one network to another in cyberspace required more or less advanced computing skills or, at the very least, a long, painful apprenticeship. This situation is now changing.

New generations of software and automated search engines are relieving Web surfers of the need to manipulate esoteric codes or make lengthy detours in their search for information. Having made the human-computer interface more user-friendly, having opened up the work space between different software programs, having simplified connections between computers and printers, scanners, and devices for capturing and regenerating sounds and images, advances in interface design are now beginning to address the opacity of cyberspace.

Advanced cellular phones, digital TV, PDAs—all these cyberspace terminals will one day have significant processing power and memory. Operating systems in these devices will integrate the tools for navigating within an increasingly transparent cyberspace.

At present, powerful programs are capable of automatically locating information and texts in databases and libraries throughout cyberspace. It is also possible to train specialized software agents, called *knowbots*, to search for information in cyberspace on a regular basis, which they can automatically present to the user in the form of an interactive structured "magazine" or personalized hyperdocument.

Gopher is a specialized software application that can be used to provide users with a kind of intelligent map and lead them to sites highlighted on the map. An interconnected system like the World Wide Web, with its various search engines, is designed to transform the Internet into a giant hypertext, independent of the physical location of computer files. Every element of information on the Web contains pointers, or links, we can follow to access other documents on related subjects. Using keyword searches, we can also use the Web to access documents located on hundreds of different computers throughout the world, as if these documents were all located in the same database or on the same hard drive.

Virtually, all texts are part of a single hypertext, a single fluid textual reservoir. We can just as easily analyze and identify images, which virtually constitute a single, limitless, and expanding hypericon, kaleidoscopic and chimerical. Rising from banks of sound effects and sampled timbres, synthesized by programs for generating sounds, automatically sequenced and arranged, streams of music will compose an inaudible polyphony, flowing together into the symphony of babel. Research on navigation interfaces is oriented, directly or indirectly, by the eventual possibility of transforming cyberspace into a single and unique virtual world, one that is immense, infinitely varied, and perpetually changing.

PART II
THEORETICAL ISSUES

CHAPTER 6

The Universal without Totality

The Essence of Cyberculture

With every passing minute there are new subscribers on the Internet, new computers interconnecting, new information sent through the network. As cyberspace grows it becomes more "universal" and the world of information less totalizable. The universality of cyberspace lacks any center or guidelines. It is empty, without any particular content. Or rather, it accepts all content, since it can connect any point with any other, regardless of the semantic load of the entities so related. I don't wish to imply that the universality of cyberspace is neutral or inconsequential, since the very process of generalized interconnection has already had—and will continue to have—immense repercussions on economic, political, and cultural life. Digitally mediated universality is transforming the conditions of life in society. However, it is an indeterminate universal, which tends to maintain its indeterminacy because each new node on the expanding network can become a producer or transmitter of new and unpredictable information and can reorganize segments of the network for its own use.

Cyberspace is a system of systems, but by this very fact, it is also a *chaotic system*. The maximum embodiment of technical transparency, through its irrepressible activity, it shelters opacities of meaning. Cyberspace ceaselessly redefines the outlines of a mobile and expanding labyrinth that can't be mapped, a universal labyrinth beyond Daedalus's wildest dreams. I refer to this universality without any centralized meaning, this system of disorder and labyrinthine transparency, as the *universal without totality*. It constitutes the paradoxical essence of cyberculture.

I would like to begin this chapter with an overview of the technical developments that serve as the basis for this key characteristic of emerging civilization and an analysis of the new pragmatics of communication inaugurated by cyberspace. I will then provide a theoretical explanation of the concept of the universal without totality. In the following chapters, I will show how the concept enables us to better understand the social movement embodied by cyberculture, its aesthetic forms, and its relationship to knowledge. I will conclude with an in-depth discussion of the urban and political issues raised by the development of cyberculture: how can we articulate the virtuality of cyberspace and the territoriality of the city?

The Technical Basis of Universality

In chapter 1, on the so-called impact of new technologies, I suggested that it is impossible to assign a fixed human meaning to a technical galaxy that is constantly changing. The cultural and social implications of digital technology are becoming more profound and varied with each new interface, with each increase in computing power and capacity, with each new connection to other technological systems. However, within this fluctuating variety, the velocity of change is paradoxically invariable. The virtualization of information and communication is a key aspect of cyberspace, which I'll discuss in the following chapters. Another immutable feature of cyberspace appears to be its tendency to "create a system," a tension toward the universal. Even if it is only at the level of the technical infrastructure, the producers of operating systems (Windows, Linux, Mac OS), programming languages (C or Java), and applications software (Word, Netscape Navigator) generally hope that their products will become, and remain, the "standard." A software application is considered a standard when it becomes overwhelmingly popular on a global basis for a specific use (to manage a computer's resources, program interactive applications on the Internet, write, navigate the

Web, etc.). Consequently cyberspace is similar to certain ecological systems: eventually a specific niche is unable to maintain a large number of competing species. The initial variety disappears, generally to the benefit of the dominant life-forms. Even when many brands coexist, the underlying *technological principles* will sooner or later adhere to a small number of international standards. As digital technology becomes increasingly important as the medium for communication and collaboration, this trend toward universalization will become more pronounced in information technology. Digital documents will circulate from machine to machine, organizational system to organizational system. Anyone who has a computer will be able to communicate with another computer anywhere on the planet. Incompatible technologies will eventually be eliminated by the market, that is, increasingly, by the end users of their products. To claim that a successful innovation is the one that succeeds in "becoming a system" with the rest of the technological environment is almost tautological.

Whatever its avatars may be in the future, we can predict that all the elements of cyberspace will continue to advance toward integration, interconnection, and the establishment of increasingly interdependent systems that are universal and transparent. This trait characterizes a number of contemporary technological systems, such as aviation, automotive engineering, and the production and distribution of electricity.[1] Nonetheless cyberspace tends toward universality and systematicity (interoperability, transparency, irreversibility of strategic choices) in a more profound sense than the other major technological systems for at least two reasons.

First, it serves as the communications infrastructure and basis of coordination for other large technological systems. Moreover, it makes possible the evolution toward universalization and the functional, organizational, and operational consistency of other systems. The development of digitization is systematizing and universalizing not only in itself but also, in a secondary sense, in enabling other technosocial phenomena that tend toward global integration: finance, trade, scientific research, media, transport, industrial production.

Second, the ultimate signification of the network, the value embodied in cyberculture, is precisely its universality. This medium tends toward the generalized interconnection of people, machines, and information. And therefore if the "medium is the message," as McLuhan claims, then the message of this medium is the universal, a transparent and unlimited systematicity, one that effectively corresponds to its designers' intentions and its users' expectations.

Writing and the Totalizing Universal

To fully understand the contemporary change in civilization, we need to examine the first major transformation in the ecology of media: the transition from oral to literate culture. The emergence of cyberspace will most likely have—already has had—as radical an effect on the pragmatics of communication as the discovery of writing.

In oral societies, linguistic messages were always received when and where they were delivered. Transmitters and receivers shared the identical situation and, most of the time, a similar universe of signification. Communication participants evolved within the same semantic bath, the same context, the same living stream of interaction.

Writing opened up a communications space that was unknown to oral societies, in which it became possible to read messages produced by people thousands of miles away, or who had been dead for centuries, or who could express themselves well in spite of significant cultural or social differences. Now participants in a communication no longer had to share the same situation; they no longer interacted directly with one another.

Independent of their conditions of transmission and reception, written messages exist "out of context." This sense of being "out of context"—initially associated only with the ecology of media and the pragmatics of communication—has been legitimated, sublimated, and interiorized by culture. It was to become the core of a certain kind of rationality and would finally lead to the concept of universality.

However, it is difficult to understand a message outside the living context in which it was produced. It is for this reason that on the receiving side, the arts of interpretation, translation, and an entire linguistic technology (grammar, dictionaries, etc.) were invented. On the transmission side, efforts were made to compose messages that could circulate anywhere, independent of the conditions of production, and which contained the keys to their interpretation within themselves, their "meaning." The idea of the universal corresponds to this practical effort. In principle, there is no need to make use of a living witness, an outside authority, the customs or elements of a particular cultural environment, to understand and acknowledge, for example, the propositions that appear in Euclid's *Elements*. The text comprises the definitions and axioms from which its theorems necessarily follow. The *Elements* is one of the best examples of this type of self-sufficient, self-explanatory message, enveloping its own meaning, which would have little pertinence in an oral society.

Classical philosophy and science, each in their own way, strive for universality. The reason is that they can't be separated from the communications system initiated by writing. The "universal" religions (not only monotheistic religions but Buddhism as well) are all based on texts. If I decide to convert to Islam, I can do it in Paris, New York, or Mecca. But if I want to practice the Bororo religion (assuming such a project has meaning), I have no alternative but to live with the Bororo. The rituals, myths, beliefs, and way of life of the Bororo are not "universal" but contextual and local. They are not based on a relationship to written texts. Obviously this does not imply an ethnocentric value judgment: a Bororo myth is part of humanity's heritage and can be virtually moving to any thinking being. Particularist religions also have their texts: writing doesn't automatically *determine* the universal; it *conditions* it (there is no universality without writing).

Like scientific or philosophic texts, which must make sense in and of themselves, contain their own justification and their own conditions of interpretation, the great texts of universalizing religion envelop the source of their authority in their very construction. The origin of religious truth is revelation. Yet the Torah, the Gospels, and the Koran *are revelation itself*, or the authentic narrative of revelation. Their discourse is no longer transmitted through a tradition that assumes its authority from the past, its ancestors, or the shared evidence of a culture. The text alone (revelation) is the basis for truth and thus escapes any conditioning context. Because truth is based on a revelatory text, the religions of the book freed themselves from dependence on a specific environment and became universal.

It is worth noting that it is the author (typical of literate cultures) who is the original *source of authority*, whereas the interpreter (the central figure in oral traditions) only actualizes or modulates an external authority. Through writing, demiurgic authors invented the self-establishment of truth.

Within the universality established by writing, it is meaning that remains unchanged by interpretation, translation, transference, diffusion, conservation. The signification of the message must be the same in all places and at all times. This universal is inseparable from an attempt at semantic closure. The effort of totalization struggles against the open-ended plurality of contexts crisscrossed by messages, against the diversity of the communities that enable them to circulate. The invention of writing is followed by the special requirements found in the decontextualization of discourse. Since then, the all-encompassing mastery of signification, the pretense of embodying "everything," the attempt to introduce the same meaning into every environment (or, for science, the same degree of accuracy), have, for us, become associated with the universal.

The Mass Media and Totality

The mass media—the press, radio, cinema, and television—at least in their classic configuration, continue the cultural extension of the totalizing universal initiated by writing. Because the media message will be read, listened to, and watched by thousands or millions of people around the world, it is composed in such a way that it is compatible with the mental "common denominator" of its recipients, the lowest interpretative capability. This is not the place to develop everything that distinguishes the cultural effects of electronic media from print. I simply want to point out a similarity. Circulating within a private space of inter-action, the media message is unable to exploit the particular context in which the receiver evolves; it neglects her singularity, social adhesion, microculture, her precise situation at a specific moment in time. It is this structure, at once reductive and overwhelming, that manufactures the undifferentiated "public" of "mass" media. By their very nature, con-temporary media, because they are reduced to the most "universal" forms of emotional and cognitive attraction, are totalizing. This is also the case, although it is much more violently expressed, with the propaganda of twentieth-century totalitarianism: fascism, Nazism, and Stalinism.

Yet there is a second, complementary aspect of electronic media such as radio and television. The decontextualization discussed earlier paradoxically initiates a different, holistic context, quasi-tribal, but on a much greater scale than in oral societies. Television, by interacting with other media, gives rise to a plane of emotional existence that reunites the members of society in a kind of fluctuating, amnesiac, rapidly evolving macrocontext. This is most obvious in "live" performance and generally whenever there is "hot" or breaking news. It was Marshal McLuhan who first described the nature of media-centric societies. The principal dif-ference between the media-centric context and the oral context is that television viewers, although they are emotionally involved in the spec-tacle, can never be practically involved. By design they are never actors in the media-centric plane of existence.

The true break with the pragmatics of communication brought about by writing can't take place with radio or TV because there is sim-ply no place within these instruments of mass distribution for true reci-procity or nonhierarchical interaction among the participants. Rather than giving rise to living interactions among one or more communities, the global context created by the media remains out of reach to those who remain its passive, isolated receivers.

The Complexity of Totalization

Many cultural forms derived from writing tend toward universality, but each of them totalizes on a different attractor: universal religions act on meaning, philosophy (including political philosophy) on reason, science on reproducible accuracy (facts), the media on capturing an audience through the heady spectacle known as "communication." In all cases, totalization operates on the identity of signification. On the plane of reality they invent, these cultural machines attempt, each in its own way, to merge the collectives they assemble with one another. Universal? A kind of virtual here and now of humanity. Yet although they manage to create some form of union through their actions, these universality-generating machines also *decompose* a multitude of contextual micrototalities: paganisms, opinions, traditions, empirical knowledge, community and craft histories. This destruction of locality is itself imperfect and ambiguous, since the products of these universal machines are in turn almost always phagocytic, relocalized, mixed with the particularities they would like to transcend. Although the universal and totalization (totalization as semantic closure, the unity of reason, the reduction to a common denominator) have always been linked, their conjunction harbors significant tensions and painful contradictions, which the new media ecology polarized by cyberspace may enable us to disentangle. It should be emphasized that this separation is in no way guaranteed or automatic. What the ecology of the technologies of communication proposes, humanity disposes. Human actors will ultimately decide, deliberately or in the semiconsciousness of collective effects, the nature of the cultural universal they build together. It is essential that they have a glimpse of the new choices available to them.

Cyberculture: The Universal without Totality
The most important cultural event engendered by the emergence of cyberspace is the disengagement of the two social operators or abstract (far more so than any concept) machinery known as universality and totalization. The reason is simple: cyberspace dissolves the pragmatics of communication, which, since the invention of writing, has conjoined the universal and totality. It brings us back to a preliterate situation—but on another level and in another orbit—to the extent that the real-time interconnection and dynamism of on-line memory once again create a

shared context, the same immense living hypertext for the participants in a communication. Regardless of the message, it is connected to other messages, comments, and constantly evolving glosses, to other interested persons, to forums where it can be debated here and now. Any text can become the fragment that passes unnoticed through the moving hypertext that surrounds it, connects it to other texts, and serves as a mediator or medium for reciprocal, interactive, uninterrupted communication. In the classical regime of writing, the reader is condemned to reactualize the context at great expense, or submit to the determined efforts of churches, institutions, or schools to revive and enclose meaning. Today, technically, because of the imminent networking of all the machines on the planet, there are almost no messages "out of context," separated from an active community. Virtually, all messages are plunged into a communicational bath that is teeming with life, including humans themselves, and cyberspace will gradually emerge as its heart.

Mail, the telephone, newspapers and magazines, book publishing, radio, and the endless number of television stations now form an imperfect fringe, the partial and unequal appendixes of an open space of interconnection animated by nonhierarchical communication, which is chaotic, tumultuous, fractal, set in motion by magmatic processes of collective intelligence. While it is true that we never step twice into the same informational stream, the density of the links and the speed of circulation are such that those involved in communication have little difficulty in sharing the same context, even if this situation is somewhat fluid and frequently confused.

Generalized interconnection, the minimal utopia and prime mover of the growth of the Internet, has emerged as a new form of the universal. But we need to be cautious. The ongoing process of global interconnection will indeed realize a form of the universal, but the situation is not the same as that for static writing. In this case the universal is no longer articulated around a semantic closure brought about by decontextualization. Quite the contrary. This universal does not totalize through meaning; it unites us through contact and general interaction.

There may be a temptation to believe that cyberculture isn't, strictly speaking, universal but planetary because of the raw geographic facts: the extension of networks for material and data transport, and the exponential growth of cyberspace. To make matters worse, doesn't the universal simply mask global phenomena, such as the globalization of the economy or financial markets?

Certainly this new universal contains a strong dose of the global and planetary, but it is not limited to them. The "universal by contact" is still universal in the most profound sense, because it is inseparable from the idea of humanity. Even the harshest critics of cyberspace acknowledge this when they lament, and rightfully, that the majority of the world's population is excluded or that Africa plays such a small role in cyberspace's development. What does the desire for "universal access" really imply? It shows that participation in this space, which connects human beings to one another, which enables communities to communicate among and with themselves, which eliminates monopolies of distribution and enables each of us to distribute what we want to whomever we want, is a right and that construction of this space appears to be a kind of moral imperative.

Cyberculture gives shape to a new form of the universal: the universal without totality. Filled with the resonance of Enlightenment philosophy, its universality stems from the fact that it maintains a profound relationship with the idea of humanity. Cyberspace engenders a culture of the universal not because it is *in fact* everywhere but because the form or idea of cyberspace implicates all human beings *by right*.

By means of computers and networks, people from all walks of life can contact one another, shake hands around the world. Rather than being constructed around the identity of meaning, this new universal is characterized by immersion. We are all in the same bath, the same communicational deluge. The question of semantic closure or totalization is no longer relevant.

A new media ecology has taken shape along with the growth of cyberspace. I can now claim, without being paradoxical, that the more universal (larger, interconnected, interactive) it is, the less totalizable it becomes. Each additional connection adds heterogeneity, new information sources, new perspectives, so that global meaning becomes increasingly difficult to read, or circumscribe, or enclose, or control. This universal provides access to a joyous participation in the global, to the actual collective intelligence of the species. Through it we participate more intensely in our living humanity. Yet our participation does not contradict the multiplication of singularity and the growth of disorder.

The more the new universal is concretized or actualized, the less totalizable it is. It's tempting to say that this is, indeed, the true universal, because it can no longer be confused with an expansion of the local or the forced export of products from a particular culture. Anarchy? Disorder? No. Those words reflect only a nostalgia for closure. By agreeing to give up a certain kind of control, we give ourselves the opportunity to encounter the real. Cyberspace is not disordered; it expresses human diversity. That maps and navigational instruments are needed for this new ocean is something on which we can all agree. But there is no a priori need to fix, to structure, to pave over a landscape that is by nature fluid and varied: an excessive desire for control will have no lasting effect on cyberspace. Such attempts at closure are becoming impossible in a practical sense or too obviously unfair.

Why bother to invent a "universal without totality" when we already have the rich concept of postmodernity at our disposal? Because they are not the same. Postmodern philosophy has accurately described the breakup of totalization. The fable of inevitable linear progress is no longer current, not in art or politics, or any other field, for that matter. When there is no longer "a" meaning to history but a multitude of small proposals fighting for legitimacy, how can we ensure the coherence of events? Where is the avant-garde? Who is ahead? Who is a progressive? To borrow an expression from Jean-François Lyotard, postmodernity has proclaimed the end of the "great totalizing narratives." Indeed, the multiplicity and radical integration of epochs, viewpoints, and legitimacies—the distinctive characteristic of the postmodern—are clearly accentuated and encouraged in cyberculture. But postmodern philosophy has confused the universal with totalization. It made the mistake of throwing out the baby of the universal with the bathwater of totality.

What is the universal? It is humanity's (virtual) presence to itself. And totality? We can define it as the stabilized collection of meanings of a plurality (discourse, situation, events, system, etc.). This global identity can be restricted to the confines of a complex process, result from the dynamic disequilibrium of life, or emerge from the oscillations and contradictions of thought. But regardless of the complexity of its modalities, totality is always the same. However, cyberculture has shown that there is another way of bringing humanity (the universal) face-to-face with its virtual presence other than through the identity of meaning (totality).

CHAPTER 7

The Social Movement of Cyberculture

It may appear strange to speak of a "social movement" for a phenomenon that is customarily thought to be technological. However, I am convinced that the emergence of cyberspace is the result of an authentic social movement, with its leaders (young, urban, and educated), its buzzwords (interconnection, virtual communities, collective intelligence), and a set of coherent goals.

Technology and Collective Desire

Even putting aside the notion of a social movement for the moment, the existence of occasionally close relationships between certain techno-industrial developments and cultural trends or phenomena of collective thinking seems somewhat obvious. The history of the automobile is particularly telling in this regard. The impressive development of the automobile over the past century, along with its consequences on the structure of our highways and cities, demography, sound and atmospheric pollution, cannot be attributed solely to the automobile industry and petroleum multinationals. The automobile responded to an immense need for autonomy and individual power. It has been invested with our fantasies, emotions, enjoyment, and frustration. The dense network of dealers and service stations, related industries, clubs, magazines, competitive events, and the whole mythology of the highway constitute a practical and mental universe into which millions and millions of individuals have poured their feelings. If it hadn't come into contact with the desires of a receptive audience that responded to its potential and brought it to life, the automobile industry on its own would never have been able to give rise to this universe. Desire served as the engine of growth. Our economic and institutional structures gave shape to this desire, channeled, refined, and inevitably diverted or transformed it.

The Infrastructure Is Not the System

If the growth of the automobile, which characterizes the twentieth century, corresponds primarily to a desire for individual power, the growth of cyberspace corresponds to a desire for reciprocal communication and collective intelligence. Yet we frequently confuse the electronic highway with cyberspace. Cyberspace is not a specific technological infrastructure for telecommunication but a way of making use of existing infrastructures, as imperfect and disparate as they are. The electronic highway refers to a group of software standards, copper and fiber-optic cables, satellite links, et cetera. Cyberspace, however, through the use of physical links, attempts to build a particular type of relationship among people. A historical analogy will help illustrate this important point. The material and organizational technologies of the postal relay system using mounted couriers existed in China since high antiquity. In use in the Roman empire as well, they were forgotten in Europe during the early Middle Ages. The relay system was borrowed from China by the vast Mongol empire during the thirteenth century.[1] The people of the steppes transmitted the principles of the system to the West, where they were lost for centuries. Beginning in the fifteenth century, some European states established a postal network for the central government. These communications networks were used to receive breaking news from all points in the kingdom and dispatch orders as quickly as possible. Early mail networks had been used for precisely this purpose in the Roman empire and in China. However, the real social innovation, the one that affected relationships among individuals, didn't arrive until the seventeenth century, with the use of postal technology for distributing mail from point to point, distant individual to distant individual, and not only from the center to the periphery and from the periphery to the center. This evolution resulted from a social need that gradually overwhelmed the original system linking the periphery to the center. At first this manifested itself in a "repurposing" of the system and other vaguely illicit activities (activities that were tolerated and sometimes encouraged by the state). These became increasingly open and were finally officially acknowledged. The subsequent system gave rise to economic and administrative correspondence, epistolary literature, the European republic of the mind (networks of scholars and philosophers), and love letters. The post office, as a social system of communication, is intimately linked to the rise of ideas and practices that enhanced freedom of expression and promoted the concept of a free contract among individuals. This example clearly demonstrates how a communications infrastructure can be taken over by a cultural movement capable of transforming its social significance and stimulating its technological and organizational development. It is worth noting that as soon as the mail system became a public service rather than a state monopoly, it tended to become a profitable economic activity, operated by private entrepreneurs. It wasn't until the nineteenth century, however, that the system would be extended to the entire European population, especially in rural areas. As a technological infrastructure, the relay system had existed for centuries, but with the inception of the new practice of unrestricted correspondence among individuals as a normal activity, classical Europe helped civilize it, invested it with a profound human significance.

Cyberspace and Social Movement

In a similar vein, California's Computers for the People wanted to put the power of computers in the hands of individuals while freeing them from the need to be supervised by technicians. The practical result of this "utopian" movement, active at the end of the seventies, was to make computers affordable to private individuals; neophytes could now learn how to use them without the need to become specialists in information technology. The social significance of information technology was transformed from top to bottom. While there is little doubt that the aspirations of the original movement were absorbed by the computer industry, we need to recognize that the industry succeeded in realizing, in its own way, the objectives of the movement. Recall that personal computing was not brought about, much less foreseen, through the efforts of a government or multinational. Its creator and principal engine was a social movement that wanted to reappropriate technological power on behalf of individuals, power that had until then been monopolized by large bureaucratic institutions.

The growth of computerized communication was initiated by an international movement of young, cultivated urbanites, which came into being at the end of the eighties. Participants in the movement explored and constructed a space for meeting, sharing, and creating in concert. Although the Internet may be the great ocean on the new informational planet, we shouldn't overlook the many small streams that feed it: corporate networks, associations, universities, as well as conventional media (libraries, museums, newspapers, television, etc.). It is this entire "hydrographic network," down to the smallest BBS,[2] that constitutes cyberspace, not only the Internet.

The people who promoted the growth of cyberspace are primarily anonymous, volunteers intent on improving the software tools used for communication rather than the cybercelebrities, government leaders, and CEOs whose names saturate the media. We tend to forget the early visionaries of cyberspace such as Douglas Engelbart and J. C. R. Licklider, who, by the sixties, believed we could use computer networks for collective intelligence; the technicians who developed the first e-mail systems and mailing lists; the students who developed, distributed, and improved communications software; and the thousands of BBS users and administrators. The symbol and principal flower of cyberspace, the Internet is one of the most amazing examples of cooperative construction on an international scale, the technical expression of a movement that developed from the bottom up and was constantly nurtured by a multitude of local initiatives.

Just as interpersonal correspondence revealed the "true" function of the postal system, so this new social movement helped develop the "true" function of telephone networks and the personal computer: cyberspace as a form of mutual interactive communication, within and among communities; cyberspace as the horizon of the living, heterogeneous, untotalizable virtual world in which every human being can share. Any attempt to confine new communications technologies to earlier media forms (one-to-many distribution methods from a transmitting center to a receiving periphery) can only restrict the scope of cyberspace for the development of civilization, even though the political and economic interests at stake are—unfortunately—perfectly understandable.

The exponential growth of Internet subscriptions at the end of the eighties clearly predates the many industrial "multimedia" projects, as well as political slogans such as the "information superhighway," which unleashed the Internet craze at the start of the nineties. Such official projects represent attempts at control by governments, the major industries, and the media of an emerging cyberspace whose true producers have invented—often intentionally—a fragile, threatened civilization unlike anything that existed in the past.

Interconnectivity

From simple to complex, three principles have helped guide the initial growth of cyberspace: interconnectivity, the creation of virtual communities, and collective intelligence.

One of the ideas, or rather one of the strongest forces behind the development of cyberspace, is that of *interconnectivity*. For cyberculture, being connected is always preferable to being isolated. Connectivity is a good in itself. Christian Huitema expressed this well when he wrote that the technical horizon of the cyberculture movement is universal communication: every computer on the planet, every device, every machine, from the automobile to the toaster, *must* have an Internet address.[3] This is cyberculture's categorical imperative. If this program is realistic, the least artifact will be able to receive and transmit information with all others, preferably over a wireless connection. Coupled with the growth in transmission capacity, the trend toward interconnectivity provokes a mutation in the physics of communication: from the concept of the channel and the network to a sensation of all-encompassing space. Information vehicles will no longer be *in* space, but through a form of topological reversal, all space will become an interactive channel. Cyberculture points toward a civilization of generalized telepresence. Above and beyond a physics of communication, interconnectivity constitutes a continuum of borderless humanity, hollows out an oceanic informational environment, immerses beings and things in the same bath of interactive communication. Interconnectivity weaves a universal through contact.

Virtual Communities

The second principle of cyberculture obviously extends the first, since the development of virtual communities is based on interconnectivity. A virtual community is constructed from related interests and knowledge, shared projects, a process of cooperation and exchange, independent of geographic proximity or institutional affiliations.

It should be pointed out for those who are unfamiliar with modern technology that on-line communication does not exclude strong emotions. Individual judgment and public opinion do not disappear in cyberspace. It is also highly unusual for networked communication to serve as a simple substitute for physical encounters. In most cases it is complementary.

Even if the influx of new arrivals dilutes it occasionally, participants in virtual communities have developed a strong sense of social morality, a set of customary, though unwritten, laws that govern their relations. This "netiquette" is primarily concerned with the pertinence of information. For example, we shouldn't post a message about one subject in a forum devoted to another. It is generally recommended that we check the forum's archive before posting messages, and, in particular, that we not post public questions whose answers are already available. Advertising is not only not recommended but generally strongly discouraged. Clearly such rules are primarily designed to avoid wasting other people's time. The implicit morality of the virtual community is generally one of reciprocity. If we learn something by reading the messages exchanged, we are also expected to provide information whenever it could be of use to someone else. The (symbolic) payback arises from the long-term reputation we develop in the virtual community. Personal attacks or insulting statements toward certain categories of individuals (nationality, sex, age, profession, etc.) are generally unacceptable. Those who do so repeatedly can be excluded by the system administrators at the request of the forum's moderators. With certain exceptions, freedom of speech is encouraged, and users are uniformly opposed to any form of censorship.

The life of a virtual community is rarely free from conflicts. These can be expressed rather brutally in the form of rhetorical diatribes known as *flames*, during which several members can "flame" anyone who disrupts the group's moral rules. Conversely, affinities, intellectual alliances, and even friendships can develop within these groups, just as they do between people who meet regularly with one another for conversation. For the participants, other members of virtual communities are indeed human, for their style of writing, skills, and opinions allow their personalities to shine through.

Manipulation and deceit are always possible in a virtual community—as they are anywhere else or in any other medium: on television, in the press, over the telephone, by mail, or during any "flesh-and-blood" meeting.

The majority of virtual communities organize the signed expression of their members in the presence of attentive readers capable of responding in the presence of other attentive readers. Consequently, as I suggested earlier, far from encouraging the irresponsibility associated with anonymity, virtual communities explore new forms of public opinion. We know that the destiny of public opinion is intimately associated with that of modern democracy. The sphere of public debate emerged in Europe during the eighteenth century with the technical support of the printing press and newspapers. In the twentieth century, the radio (especially during the thirties and forties) and television (beginning in the sixties) displaced, amplified, and confiscated—all at the same time—the exercise of public opinion. Is it therefore unreasonable to anticipate a new metamorphosis, a new complication in the very concept of "public," when virtual communities in cyberspace provide collective debate with a venue that is more open, participatory, and distributed than that of our classic media?

Virtual relationships are not a simple substitute for physical inter-action or travel, which they often help promote. In general it is a mistake to interpret the relationship between old and new methods of communication in terms of substitution. I will discuss this at greater length in a subsequent chapter, but I would first like to outline the principal arguments in support of my claim. The cinema did not eliminate the theater; it displaced it. We speak as much as we did before we learned to write, but differently. Love letters do not prevent lovers from kissing. Individuals who make the most phone calls are also those who have the greatest number of physical meetings with others. The development of virtual communities accompanies the general development of contacts and interactions of all kinds. The image of the isolated loner in front of a computer screen is based more on fantasy than on sociological inquiry. In reality, Internet subscribers (students, researchers, teachers, traveling salesmen, writers and intellectuals, etc.) probably travel more than the average person. The only drop in airport use during the last few years occurred during the Gulf War; the extension of cyberspace had nothing to do with it. On the contrary, from a global perspective, communication and transport have grown in tandem throughout the century. We shouldn't be deceived by words. A virtual community is not unreal, imaginary, or illusory. It is simply a more or less permanent collective organized around the new global electronic mail system.

Aficionados of Mexican cuisine, Angora cat lovers, C fanatics, and interpreters of Heidegger, who were once scattered across the globe, often isolated or, at the very least, without regular contact among them-selves, now have access to a place where they can meet and exchange ideas on a regular basis. It's reasonable to claim, therefore, that such virtual communities bring about a true actualization (in the sense of effectively putting people in contact) of human groups, groups that were merely potential before the arrival of cyberspace. It would be far more accurate to use the term "actual community" to describe the phenomena characteristic of collective communication in cyberspace.[4]

Along with cyberculture arises the desire to construct a social bond that is based not on territorial or institutional affiliations, or relationships of power, but on common interests, games, shared knowledge, cooperative apprenticeship, open processes of collaboration. Our desire for virtual communities reflects an ideal of deterritorialized human relationship, nonhierarchical and free. Virtual communities are the engines and agents, the multiform and astonishing life-forms of the universal through contact.

Collective Intelligence

Any human group can form a virtual community. Such communities are valuable to the extent that they approach the ideal of an intelligent collective, one that is more imaginative, responsive, and capable of learning and creating than an intelligently directed collective. Cyberspace may simply be the necessary technological detour through which we reach collective intelligence.

The third principle of cyberspace, collective intelligence is its spiritual aspect, its ultimate goal. This project was carried forward by the visionaries of the sixties: Engelbart (inventor of the mouse and windows), Licklider (newsgroup pioneer), and Ted Nelson (inventor of hypertext). The ideal of collective intelligence is also manifested by some of the "gurus" of today's cyberculture such as Tim Berners Lee (inventor of the World Wide Web), John Perry Barlow (former lyricist and songwriter for the Grateful Dead and one of the founders and spokesman of the Electronic Frontier Foundation), and Marc Pesce (coordinator of the VRML standard). Collective intelligence has also been developed by the commentators and philosophers of cyberculture such as Kevin Kelly, Joël de Rosnay, and myself.[5] It is also practiced on-line by a growing number of netsurfers, newsgroup participants, and virtual communities of all kinds.

Collective intelligence is more a field of problems than a solution. It is commonly recognized that the best use we can make of cyberspace is to combine the experience, imagination, and spiritual energy of those who are connected to it. But in what sense? Based on what model? Are we building human hives or anthills? Should each network give birth to some big, collective animal? Or on the contrary, is it our goal to enhance the personal contributions of everyone connected and put the resources of the group at the service of individuals? Is collective intelligence an efficient mode of coordination, in which each of us can be considered a center? Or do we want to subordinate individuals to an organism that is larger than any of them? Is the intelligent collective dynamic, autonomous, emergent, fractal? Or well defined and controlled by some supervisory organization? Will each of us become a neuron in a planetary megabrain, or will we constitute a multitude of virtual communities in which nomadic brains join together to produce and share meaning? These alternatives, which only partially intersect, define some of the fault lines that divide the project and practice of collective intelligence from within.

The extension of cyberspace will transform the constraints that have dictated the range of possible solutions to political philosophy, the management sciences, and our organizational traditions in general. A number of these constraints have disappeared with the availability of new tools of communication and coordination. We can now envisage radically new ways of organizing human groups and relationships between individuals and collectives, which lack both historical models and precursors in animal societies. I again want to emphasize that collective intelligence, whose ambivalence I have alluded to, presents an open field of problems and practical research.

A Program without Goals or Content

The social and cultural movement that underlies cyberspace—a powerful and increasingly broad movement—is converging not toward any particular content but toward a form of communication that is unmediated, interactive, community based, nonhierarchical, and rhizomatic. Generalized interconnection, the hunger for virtual communities, the exaltation of collective intelligence—none of these constitute the elements of a political or cultural program in the classic sense of the term. Yet all three are secretly driven by two essential "values": autonomy and an openness toward alterity.

Interconnectivity for the purpose of interactivity is supposed to be beneficial, regardless of the hardware, individuals involved, and places or times when we connect. Virtual communities are said to be an excellent means (among a hundred others) of making a society, whether their goals are ludic, economic, or intellectual, whether their interests are serious, frivolous, or scandalous. Collective intelligence is presumably humanity's mode of fulfillment, one that is fortunately promoted by the universal digital network, although we have no a priori knowledge of the results toward which the organizations that are merging their intellectual resources are working.

In short, the cyberculture program is the universal without totality. Universal because interconnectivity must not only be global but achieve compatibility or general interoperability. Universal because, ideally, anyone will be able to access diverse virtual communities and their products from anywhere. Universal because the program of collective intelligence can be applied to companies just as it can to schools, to geographic regions as well as international associations. Cyberspace is the tool for organizing communities of all sizes and kinds into intelligent collectives, as well as an instrument that enables intelligent collectives to structure themselves. The same software and hardware tools are now used to support the internal and external policies of collective intelligence: intranets and the Internet.[6]

General interconnectivity, virtual communities, collective intelligence—they are all figures of a universal through contact, a universal that grows like a population, which puts forth shoots, a universal that spreads like ivy.

Each of these three figures serves as the necessary condition for the subsequent figure: there can be no virtual community without interconnectivity, no large-scale collective intelligence without the virtualization or deterritorialization of communities in cyberspace. Interconnectivity conditions the virtual community, which is a potential collective intelligence.

However, these forms are initially empty. There is no external finality, no particular content to enclose or totalize the cyberculture project, which is an integral part of the unfulfilled process that incorporates inter-connectivity, the development of virtual communities, and the intensifi-cation of a fractal collective intelligence that is variously reproducible and everywhere different. The continuous movement of interconnectivity to promote interactive any-to-any communication is itself a powerful indicator that totalization will not occur, that sources will always remain heterogeneous, that mutagenic systems and vanishing lines will proliferate.

CHAPTER 8

The Sound of Cyberculture

In this and the following chapter, I would like to explore the artistic and aesthetic dimensions of cyberculture. Starting with an examination of the configurations of communication and interaction that emerge in the technosocial environment of cyberculture, I analyze the new modes of producing and receiving cultural artifacts. Artistic questions will be examined from the specific point of view of the pragmatics of creation and appreciation.

The Arts of the Virtual

There are a number of genres associated with cyberculture: automatic composition of music or text, sampling and rearrangement, artificial life and autonomous robots, virtual worlds, aesthetic or cultural Web sites, hypermedia, networked or interactive events, hybrid mixtures of the "virtual" and the "real," interactive installations.[1] In spite of this variety, it's possible to distinguish some of the key features of the art of cyberculture, which, although they are not necessarily all present in individual works, are nevertheless representative of its principal tendencies.

One of the most consistent characteristics of cyberart is the participation in the work of those who experience, interpret, explore, or read it. I'm referring to not only their participation in constructing meaning but, rather, their coproduction in the actual work. In cyberart the spectator is called upon to participate directly in the actualization (the materialization, display, publication, effective occurrence here and now) of a sequence of signs or events.

In a similar vein, the organization of the process of collective creation is also typical of the arts of the virtual and frequently involves collaboration between initiators (artists) and participants, the networking of artists working on the same project, recording the interactions or activities that ultimately constitute the work, collaboration between artists and engineers, and so forth.

Collective creation as well as audience participation go hand in hand with a third characteristic of cyberart: continuous creation. The virtual work is "open" by design. Every actualization reveals a new aspect of the work. Some systems not only manifest a combination of possibilities but encourage the emergence of absolutely unpredictable forms during the process of interaction. Thus creation is no longer limited to the moment of conception or realization; the virtual system provides a machine for generating events.

Techno finds its material in the great storehouse of sampled sound. If it weren't for the legal and financial problems that plague its producers, hypermedia would be constructed from available images and text. Computer programs can assemble "original" texts by recombining fragments of a preexisting corpus. Web sites constantly refer back and forth to one another, their hypertext structure promoting the interpenetration of messages, a mutual immersion in virtual spaces. Following the avant-gardes of the twentieth century, cyberart reformulates questions concerning the limits of art and its frame with particular acuity.

All the foregoing characteristics, the active participation of an audience, collective creation, work-as-event, work-as-process, the interconnection and blurring of limits, the work emerging—like a virtual Aphrodite—from an ocean of digital signs, help undermine (rather than eliminate) two figures that have until now ensured the integrity, substantiality, and possible totalization of works of art: the author and the archive. An art of the virtual is both possible and desirable, even if these figures fade into the background. But cyberart demands new criteria of judgment and conservation, which often conflict with the current attitudes of the art market, the training of critics, and the practices of museums. This art, which revives the tradition of games and rituals, also demands the invention of new forms of collaboration among artists, engineers, and sponsors, both public and private.

My argument can be summarized as follows: the form of the universal-without-totality, a characteristic of the civilization engendered by digital networks in general, also enables us to appreciate the specificity of the artistic genres unique to cyberculture. In this chapter, I will turn my attention to music (and especially techno and sampled music), generalizing my argument to other arts in the following chapter.

The Globalization of Music
Today's popular music is global, eclectic and changing, without any unifying system. We can immediately recognize in it certain traits characteristic of the universal-without-totality. Historically, this state is very recent. The first step toward a universal, untotalizing music was taken with the use of sound recordings and radio broadcasts. When we examine early record catalogs, which appeared at the beginning of the twentieth century, we discover a musical landscape that is far more static and compartmentalized than the one we are familiar with today. At the time, listeners weren't accustomed to music from around the world and wanted to hear what they had always heard. Every country, indeed, every region and microregion, had its singers, its songs in its own dialect, its specific melodies and instruments. Almost all popular music was recorded by local musicians for a local public. Only recordings of classical music from the written Western tradition had an international audience. Nearly a century later, the situation has radically changed, since recorded popular music is often "world" music. Moreover, it is constantly changing, since it continuously integrates local traditions along with the expression of new cultural and social currents.

Two series of intertwined mutations explain the transition from one state to the other in the international musical landscape: one involved the general transformation of the economy and society (globalization, increased travel, extension of an international urban and suburban lifestyle, cultural and social youth movements), which I won't discuss at length here, and the other involved the economic and technological conditions associated with recording, distributing, and listening to music.

The distribution of audio records resulted in a standardization of popular music that was similar to the effect that printing had on language. During the fifteenth century, in countries such as France, England, and Italy, there were as many "dialects" as there were rural microregions. But a book had to target a market that was large enough to make the book profitable. When books were printed in the vernacular and no longer exclusively in Latin, a choice had to be made among the many local languages, and *a* national language extracted. Tuscan, the dialect of Touraine, and the English of the court became Italian, French, and English, relegating—with the help of royal governments—other languages to the status of local dialects. In his translation of the Bible, Luther combined different German dialects and thus helped forge *the* German language, or written German.

For similar reasons, the evolution of popular music catalogs since the beginning of the twentieth century shows that starting with an initial fragmentation, national and international musics have been formed gradually. This mutation is especially noticeable outside the West, where urbanization and the cultural influence of a central government was still relatively limited at the beginning of the century. That music is independent of language (with the notable exception of the text for popular songs) has obviously facilitated this phenomenon of decompartmentalization. If writing decontextualizes music, its recording and reproduction create a worldwide sound context—along with the ears to listen.

As long as the quality of recordings didn't exceed a certain threshold, radio broadcast only live music. When FM stations became popular after the Second World War and began to broadcast high-quality recordings, the phenomenon of global popular music took off, primarily through rock and pop during the sixties and seventies.

Although globalization might have been thought to lead to some final homogenization, a kind of musical entropy in which styles, traditions, and differences melt together into a uniform mass, world pop music retained its distinct flavor within the global "soup." Some regions of the musical landscape—I'm thinking primarily of regions of the Third World where music is distributed on cassette—are protected or disconnected from the international market. World music continues to thrive on these imperceptible but still living isolates, ancient local traditions, as well as an unsullied and widespread poetic and musical creativity. New genres, new styles, new sounds appear all the time, re-creating the potential differences that agitate the planetary musical space.

The dynamic of world popular music is an illustration of the universal-without-totality: universal through the diffusion of musical forms and listening habits that are planetary, without totality because global styles are multiple and undergo constant transformation and renewal.

But the exemplary figure of the new universal doesn't fully appear until the arrival of digitization—techno and, more generally, all music made from digital samples are the sound of cyberculture. To get some idea of techno's originality, its process of creation and circulation, we need briefly to examine earlier modes for the transmission and renewal of music.

Oral, Written, and Recorded Music

In oral cultures, music is direct, distributed by imitation, evolves through the reinvention of timeless themes and genres. The majority of melodies have no identifiable authors; they belong to tradition. Yes, individual poets and musicians are capable of writing songs or even winning prizes or competitions. The creative role of the individual is certainly not ignored. Yet in oral cultures, the figure of the great interpreter, the one who transmits a tradition by instilling it with new life, is far more common than that of the great composer.

Writing music brings about a new form of transmission, no longer from body to body, from ear to mouth and hand to ear, but through text. Although interpretation—that is, aural actualization—is subject to a continuous process of initiation, imitation, and reinvention, written music, composition, is stationary, detached from the context in which it is received.

Western music, based on writing and a combination of neutral sounds (detached from any magical, religious, or cosmological associations), is presented as being universal and is taught as such in conservatories throughout the world. The appearance of a written tradition reinforces the figure of the composer who signs a score and claims originality. In place of the invisible drift of genres and themes that characterize oral cultures, writing conditions a *historical* evolution in which every innovation is clearly distinct from previous forms. Anyone can determine the intrinsically historical character of the Western scholarly tradition; by simply listening to a piece of music, we can date it approximately, even if we don't know the author. In concluding this aside on the effects of notation, I'd like to point out the link between static writing and the three cultural figures of universality, history, and the author.

Through writing, music that was based on an oral tradition was made to conform to a different cultural cycle. Similarly, recording established styles of interpretation for written music while it helped control their evolution. It was no longer merely the abstract structure of a piece that could be transmitted and decontextualized, but its actualization as sound as well. Recording became a means for archiving and historicizing musical forms that had remained within the orbit of oral tradition (musical ethnography). And certain musical genres, such as jazz or rock, currently exist only because of a "recording tradition."

Toward the end of the sixties, the multitrack recording studio became the great integrator, the principal instrument of musical creation. From this time onward, for a growing number of musical scores, the original was the recording made in the studio, which concert performances never succeeded in reproducing. Among the first examples of this paradoxical situation, in which the recording became the original, were tracks from the Beatles' *Sergeant Pepper* album, whose complexity required the use of mixing techniques that were impossible to duplicate in a concert setting.

Techno

Like notation and recording, digitization inaugurated a new pragmatics of creation and musical listening. Earlier I noted that the recording studio had become the principal instrument, or metainstrument, of contemporary music. One of the first effects of digitization was to make studio technology affordable to the individual musician. The major functions of the digital studio, controlled from a simple personal computer, include a sequencer for use in composition, a sampler for digitizing sound, software for mixing and arranging the digitized sound, and a synthesizer, which produces sound from a sequence of instructions or digital codes. The Musical Instruments Digital Interface (MIDI) standard can be used to "play" any sequence of musical instructions produced in a digital studio on any synthesizer. With the introduction of MP3 compression, music files can easily be exchanged over the Internet and played on the user's computer.

This meant that musicians could now personally control the entire musical production chain and even place their products on the network without going through the intermediaries that had introduced the use of notation and recording (publishers, artists, the big studios, stores). In one sense, we have returned to the simplicity and personal appropriation of musical production that characterized the oral tradition. Although the renewed autonomy of the musician is an important element in the new ecology of music, it is primarily through the dynamic of collective creation and listening that the effects of digitization appear most original.

It has become increasingly frequent for musicians to produce their music by sampling and rearranging sounds, and sometimes entire pieces, taken from the stock of available recordings. This sampled music can itself be variously sampled, mixed, and transformed by other musicians in an endless cycle, a technique that is especially common in techno music. For example, jungle uses only sampling; acid jazz is made by sampling old jazz recordings.

Techno and sampling have invented a new modality of tradition, an original way of weaving the cultural fabric. Unlike the oral or recording traditions, rehearsals and the inspiration that comes from listening to prerecorded music are no longer dominant. Nor is the sense of musical interpretation that exists between the score and its execution or the relation of reference, progression, and competitive invention that exists among composers, which typify the written tradition. In techno each actor in the creating collective selects sound material from a flux circulating within a vast technological network. This material is mixed, arranged, transformed, and reinjected as an "original" work into the stream of digital music. Every musician, every group of musicians, functions as an operator on a continuously changing flux within a cyclical network of cooperators. Never have creators been in such intimate relationship with one another as they have in this type of digital tradition. Here the link is formed by circulating the musical and sound material itself, not simply by listening, imitating, and playing.

Recording is no longer the ultimate goal or musical reference. It is merely the ephemeral trace (destined to be sampled, deformed, mixed) of a particular act within a collective process. This is not to say that recording has no importance or that techno musicians are totally indifferent to the fact that their productions refer. But it is more important to *create an event* within the circuit (for example, as part of a rave) than to add a memorable item to our musical archives.

Cyberculture is fractal. Each of its subsystems reveals a form that resembles a global configuration. In techno music we find the three principles that characterize the social movement of cyberculture outlined earlier.

Interconnectivity can obviously occur through technical standardization (MIDI, MP3) and use of the Internet, but it is also apparent in the continuous flux of sound material that circulates among musicians and the possibility of digitizing and processing any piece of music (virtual interconnectivity). Bear in mind that this circulation in a recursive sampling network, where each nodal operator helps produce a whole, is self-enhancing: it is a priori a *good form*.

Techno music is so compatible with the principle of virtual community because musical events are often produced during rave parties and assume meaning within more or less ephemeral communities of musicians and disc jockeys. Web sites used for the exchange of MP3 files are also good examples of virtual communities.

When a musician offers the community a finished work, he or she also adds to the stock from which others work. Every musician is a producer of raw material, transformer, author, interpreter, and listener in an unstable and self-organized circuit of cooperative creation and concurrent appreciation. This process of collective musical intelligence is constantly expanding and gradually integrates the entire heritage of recorded music.

Techno and, in general, all fundamentally digital music are an illustration of the singular figure of the universal-without-totality. Universality results from the compatibility or technical interoperability and the facility with which sounds circulate and reproduce themselves in cyberspace. But the universality of digital music also prolongs the musical globalization favored by the recording industry and FM radio. Ethnic, religious, classical, folk, jazz—music of all kinds is torn from its original context, sampled, mixed, transformed, and finally served up to an audience of committed listeners in a state of permanent apprenticeship. The transglobal underground genre, for example, participates intensely in the process during universalization of its content through contact and fusion. It integrates tribal or liturgical music with electronic or even industrial sounds, which incorporate hypnotic or frenetic rhythms that try to induce trancelike effects. Unlike classical Western music, which was written, the new musical universal will not introduce the same system everywhere: it distributes a universal through contact, one that is nonhierarchical, eclectic, constantly mutating. A musical flux in constant transformation gradually invents the space it enlarges. This flux is also universal to the extent that, following the advance of digitization, it is sustained by the unrestricted "wholeness" of music, from the most modern to the most archaic.

CHAPTER 9

The Art of Cyberculture

Cyberculture and Society

The canonical genre of cyberculture is the *virtual world*. This shouldn't be interpreted in the narrow sense of the term as a computerized simulation of a three-dimensional universe, which we explore with a stereoscopic headset and data gloves, but as a digital store of sensory and informational virtualities that are actualized only through interaction with human beings. Depending on the technology, this actualization can be more or less creative and unpredictable, with a variable component that is dependent on human initiative. Virtual worlds can be collectively enriched and experienced, thus becoming a meeting place and communications medium shared by their participants.

The *engineer of worlds* will be the major artist of the twenty-first century. He or she will create virtualities, design communication spaces, build the collective hardware of cognition and memory, structure sensorimotor interaction with the data universe. The World Wide Web, for example, is a virtual world that promotes collective intelligence. Its inventors—Tim Berners Lee and the people who programmed the interfaces for navigating it—were engineers of worlds. Software developers, video game designers, and the artists who explore the boundaries of interactive devices and televirtual systems are also engineers of worlds.

In general we can distinguish two major types of virtual world:

- those that are limited and editorialized, such as CD-ROMs and "closed" (off-line) installations by artists
- those that are accessible over a network and infinitely open to interaction, transformation, and connection with other virtual worlds (on-line)

There is no reason to contrast *on-line* and *off-line*, as was done in the past. They are complementary relations, feeding and inspiring each other.

Off-line works of art can conveniently provide partial and temporary projections of collective intelligence and imagination, which can be disseminated in a networked environment. They are able to take advantage of technical innovations, specifically, freedom from low bandwidth limitations. And they help constitute original or creative isolates outside the continuous flow of communication.

Similarly, virtual worlds accessible on-line can be supplied with off-line data and replenish them in turn. These are essentially interactive communications environments. But in this case, the virtual world operates as a message depot, a dynamic context accessible to everyone, a communal memory that is collectively replenished in real time.

The development of the technical infrastructure of cyberspace opens the perspective of interconnected virtual worlds. The gradual fusion of digitized texts into a single immense hypertext[1] is only the prelude to a more general form of interconnection, which will blend all digitized information, including film and interactive[2] three-dimensional environments. In this way, the network will provide access to a gigantic, heterogeneous virtual metaworld, which will embrace the eruption of particular virtual worlds and the dynamic links, interconnecting passages, corridors, and burrows that characterize this digital wonderland. The virtual metaworld, or cyberspace, will become the principal locus of communication, economic transaction, learning, and entertainment for human societies. In it we will experience the beauty stored in the memory of ancient cultures, together with the forms specific to cyberculture. Just as cinema did not replace the theater but created a new genre with its own tradition and codes, so the emerging genres of cyber-culture, like techno music and virtual worlds, won't replace earlier ones. They'll add to civilization's heritage while reorganizing the communications economy and the system of the arts. Features such as the demise of the author and the recorded archive won't affect art or culture in general, but only those works specifically associated with cyberculture.

Even off-line, the interactive work demands the involvement of those who experience it. The interactant participates in structuring the message he or she receives. As with the creations of the engineer of worlds, multiplayer virtual worlds are the collective creations of their explorers. The artistic artifacts of cyberculture are works of flow, process, and incident that do not lend themselves to archiving and conservation. In cyberspace every virtual world is potentially linked to all others, envelops them, and is contained by them according to a paradoxical topology that intertwines inside and outside. Even now many works of cyberculture art have no clear boundaries. They are "open works"[3] not only because they admit a multitude of interpretations but especially because they are physically receptive to the active immersion of an explorer and materially interconnected with other works on the network. The extent of this openness is obviously variable; however, the more the work exploits the possibilities offered by interaction, interconnection, and the tools of collective creation, the more typical it is of cyberculture and the less it behaves as a "work of art" in the traditional sense.

The cybercultural work achieves a certain kind of universality through its ubiquitous presence on the network, its connection to other works and their copresence, and through a material openness, rather than its association with some universally valid meaning. This form of universality through contact goes hand in hand with a tendency toward detotalization, for the guarantor of the totalization of the work, the closure of the meaning associated with it, is the author. Even if the meaning of a work is said to be open or multiple, an author must still be presupposed if we intend to *interpret* his or her intentions, decode a project, a social expression, or even an unconscious mind. The author is the condition of possibility for any perspective of stable meaning. Yet it has become commonplace to claim that cyberculture calls into question the importance and function of the signatory. The engineer of worlds does not sign a finite work but an environment that is essentially incomplete. It is the explorers of these worlds who will construct not only its variable, multiple, and unexpected meanings but the order of reading it and the work's perceivable forms. The continuous metamorphosis of adjacent works and the virtual environment that supports and penetrates the work will help dispossess a potential author of his or her prerogatives as a guarantor of meaning.

Fortunately, sensitivity, talent, ability, and individual creative effort are still popular. But these can apply just as well, and perhaps more appropriately, to the interpreter, the "performer," the explorer, the engineer of worlds, each member of the construction team, as they do to an increasingly nebulous author.

After the author, the second condition for totalization, or the closure of meaning, is the physical closure associated with the work's temporal fixity. The record, the archive, the artwork that can be conserved in a museum are *finished* messages. A painting, for example, an object of conservation, is both the work itself and its archive. But the work-as-event, the work-as-process, the interactive, metamorphic work, the interconnected, crisscrossed, indefinitely coconstructed work of cyberculture, is difficult to record as such, even if we photograph a moment in its process or capture a partial trace of its expression. To create a work, record, or archive no longer has—can no longer have—the same meaning as before the information deluge. When there are few archives, when they are harder to circumscribe, the process of discovery will again involve entering the flow of memory. But when memory is practically infinite, in flux, overflowing, fed each second by a myriad of sensors and millions of people, *cultural archives no longer serve to differentiate or brand.* The act of creation par excellence will then consist in creating an event, here and now, for a community, in constituting the collective in which the event will occur, in partially reorganizing the virtual metaworld, the unstable landscape of meaning that shelters us and our works.

The pragmatics of communication in cyberspace erase the two

leading elements of traditional totalization: *intentional* totalization by the author and *extensional* totalization through recording. Using the concepts of the rhizome and the plane of immanence, Deleuze and Guattari[4] have philosophically described an abstract schema that comprises

- the a priori unlimited proliferation of connections among heterogeneous nodes and the mobile multiplicity of centers within an open network
- the agitation of intertwined hierarchies, the holographic effects of the partial—and everywhere different—envelopment of wholes in their parts
- the autopoietic and self-organizing dynamic of mutant populations that extend, create, and transform a qualitatively varied space, a landscape punctuated by singularities

This schema is actualized *socially* through the life of virtual communities, *cognitively* through the process of collective intelligence, and *semiotically* as the great hypertext or virtual metaworld of the Web.

The cybercultural work participates in these rhizomes, this plane of immanence in cyberspace. It is riddled with tunnels and faults, which expose it to an unidentifiable exterior, and interconnected (or awaiting connection) to people and the flow of data.

This is the global hypertext, a virtual metaworld in perpetual metamorphosis, an abundant musical or iconic flow. Everyone is asked to become a singular operator, qualitatively different, in the transformation of the universal and untotalizable hyperdocument. A continuum extends between the visitor and the engineer of virtual worlds, between those who are content merely to visit and those who will design systems or sculpt data. This reciprocity is in no way ensured by our technological evolution; it is only a fortunate possibility exposed by the new means of communication available to us. It is up to our social actors and cultural activists to take advantage of it. In doing so, they will avoid reproducing in cyberspace the malignant dissymmetry that characterizes contemporary mass media.

The Universal without Totality: Text, Music, Image

For each great modality of signs—alphabetic texts, music, images—cyberculture produces new forms and new ways of interacting. The *text* folds, folds again, divides, and adheres to itself in bits and pieces; it mutates into hypertexts and these hypertexts connect to one another to form the indefinitely open and mobile hypertextual plane of the Web.

Music certainly lends itself to discontinuous navigation through hyperlinks (the listener can move from sound unit to sound unit), but it gains much less than text. Its most important mutation in the conversion to a digital format is characterized primarily by the recursive process opened through exchanging, sampling, mixing, and rearranging, that is, through extending a virtual musical ocean continuously fed and transformed by a community of musicians.

The image loses its exteriority as spectacle and opens itself to immersion. Representation makes room for the interactive visualization of a model; simulation replaces resemblance. Drawings, photographs, and films are voided; they welcome the active explorer of a digital model, a collective of workers or players engaged in the cooperative construction of a data universe.

We have, therefore, three principal forms in cyberspace:

- the networked read-write hyperdocument for text
- the recursive process of creation and transformation of a memory stream by a community of differentiated cooperators in the case of music
- the sensorimotor interaction with a set of data, which defines the virtual status of the image

However, none of these three forms excludes the others. What's more, each of them can actualize the same abstract structure of the untotalizable universal differently, and in one sense each contains the others.

We navigate within a virtual world as we do in a hypertext, and the pragmatics of techno also assume a principle of virtual off-line and on-line navigation in musical memory. Even some real-time musical performances implement the techniques of hypermedia.

In my analysis of new trends in digital music, I highlighted the cooperative and continuous transformation of an informational reserve that serves as both channel and shared memory. Yet this type of situation affects collective hypertexts and virtual worlds of communication as much as it does techno and MP3 files. Image and text are being increasingly subjected to sampling and rearrangement. In cyberculture every image is potentially the raw material of another image; every text can serve as the fragment of a larger text composed by an intelligent software "agent" during a specific search.

Finally, interaction and immersion, which are typical of virtual realities, illustrate a principle of the *immanence of the message for its receiver*, which can be applied to all digital modalities: the work is no longer remote but nearby. We participate in it, transform it, are its authors in part.

The immanence of messages for their receivers, their openness, the continuous and cooperative transformation of a memory stream by groups of humans—all these traits actualize the decline of totalization.

As for the new universal, it is realized in the dynamic of interconnection that typifies on-line hypermedia, participation in the mnemonic or informational ocean, the ubiquity of the virtual within the networks that carry it. Universality arises from the fact that we all bathe in the same information flow and from the loss of totality of its diluvial flood. Not content to flow forever, Heraclitus's stream has overrun its banks.

The Author in Question

Authors and archives guarantee the totalization of works of art, ensure the conditions for an all-encompassing comprehension and stable meanings. If the essence of cyberculture lies in the nontotalizing universal, we must examine, if only as a hypothesis, the guises art and culture need to assume for these two figures to fade into the background. It is unlikely, having experienced a state of civilization in which the memorable archive and creative genius have been so consequential, that we could conceive of a situation (putting aside some cultural catastrophe) where author and record have disappeared entirely. Yet we need to envisage a future state of civilization in which these two custodians of waning totalization will play only a modest role for those who produce, transmit, and experience the works of the mind.

The concept of the author in general, like the various notions of the author in particular, is strongly linked to certain forms of communication, to our social relations on the economic, legal, and institutional level. In societies where the principal mode of transmitting explicit cultural content is speech, the concept of the author is minimized and sometimes nonexistent. Myths, rites, and traditional plastic or musical forms are timeless, and we do not generally associate them with a signature, or if we do, it belongs to a mythic author. Even the concept of a signature, like that of personal style, implies writing. Artists, singers, bards, storytellers, musicians, dancers, and sculptors are generally considered to be interpreters of an immemorial theme or pattern, which is part of the heritage of the community. Within the diversity of epochs and cultures, the concept of the interpreter (along with our ability to distinguish and appreciate the great interpreters) is much more widespread than that of the author.

Obviously, the author begins to assume increasing depth with the appearance of writing. However, until the end of the Middle Ages, writers of original texts weren't necessarily considered authors. The term was reserved for an "authoritative" source, such as Aristotle; the commentator or copyist who provided the gloss was not referred to as an author. With the arrival of printing and the industrialization of text reproduction, it became necessary to precisely define the economic and legal status of writers. Consequently, the modern concept of the author developed in tandem with the notion of authorial "rights." Similarly, the Renaissance witnessed the arrival of the artist as a creative demiurge, an inventor or designer, and no longer as an artisan, a more or less inventive bearer of a tradition.

Are there any great works, great cultural creations without an author? Clearly there are. Greek mythology, for example, is one of the jewels of our cultural heritage. Yet it is unquestionably a *collective creation*, authorless, arising from an immemorial past, polished and enriched by generations of inventive retransmitters. Homer, Sophocles, and Ovid, as celebrated interpreters of this mythology, have obviously given it a certain luster. But Ovid is the author of the *Metamorphoses*, not of mythology; Sophocles wrote *Oedipus Rex*, and didn't invent the saga of the kings of Thebes.

The Bible is another example of a major work arising from humanity's spiritual and poetic past that has no distinct author. Hypertext before the fact, its construction results from a selection (sampling) and belated amalgamation of a large number of heterogeneous texts written at different times. The origin of these texts can be traced to the ancient oral traditions of the Jews (Genesis, Exodus). But it also bears the influence of Mesopotamian and Egyptian civilization (certain parts of Genesis, the Songs of Solomon) in its heightened moral reaction to a specific political and religious actuality (the books of the Prophets), its poetic and lyrical expansiveness (Psalms, Canticle of Canticles), its desire for legislative and ritual codification (Leviticus) and preservation of a historical memory (Chronicles, etc.). Yet we rightly consider the Bible to be *a* work, the bearer of a complex religious message embodying an entire cultural universe.

Within the Jewish tradition, an interpretation by a legal scholar doesn't assume its fullest authority until it becomes anonymous, when the name of the author has been erased and is integrated within a shared heritage. Talmudic scholars constantly cite the advice and comments of the wise men who preceded them, thus rendering the most precious aspect of their thought immortal in a way. Yet paradoxically the sage's greatest accomplishment consists in no longer being cited by name and disappearing as an author, so that his contribution is identified with the immemoriality of the collective tradition.

Literature isn't the only field where major works of art remain anonymous. Like the *Song of Roland*, Indian ragas, the paintings at Lascaux, the temples of Angkor, and the gothic cathedrals are unsigned.

There are great works of art without authors. Still, it appears difficult to experience beautiful works without the intervention of great interpreters, without talented individuals able to follow the thread of a tradition, reactivate it, and add to it a special brilliance. These interpreters may be well known but still remain nameless. Who was the architect of Notre-Dame de Paris? Who sculpted the portals of the cathedrals of Chartres and Rheims?

The figure of the author emerges from a media ecology and a highly specific economic, legal, and social configuration. It shouldn't be surprising that it would fade into the background when the communications system and social relations are transformed, destabilizing the cultural compost that witnessed its growth. This may be less serious than it appears, however, since the preeminence of the author determines neither the explosion of culture nor artistic creativity.

The Decline of Recording

I noted earlier that creating or recording a work, leaving a trace, no longer has the same meaning or value it had before the informational deluge. The natural result of the inflation of information is its devaluation. From that point on, the focus of the artist's work shifts to the event, that is, toward the reorganization of the landscape of meaning, which fractally inhabits all levels of the communications space, group subjectivities, and the sensible memory of individuals. Like the fabric of humanity, *something is happening* within the network of signs.

Let's be perfectly clear. This process does not involve the anticipated displacement of the resolutely material "real" preserved in museums by a labile "virtual" in cyberspace. Did the irresistible rise of the "imaginary museum" extolled by Malraux—the multiplication of art catalogs, books, and films—reduce museum attendance? On the contrary. As awareness of the recombinable elements of the imaginary museum spread, so did the number of publicly accessible facilities designed to house and expose the physical presence of art. Nonetheless if we examine the fate of a given work of art, we find that most people experience a reproduction rather than the original. Similarly, it is unlikely that virtual museums will ever compete with real museums; rather, they will serve as a kind of publicity department. They will, however, become the principal interface through which the public experiences art, just as records exposed more people to Beethoven and the Beatles than did concerts. The mistaken notion of the *substitution* of a feigned "real" by an unknown and denigrated "virtual" has resulted in a great deal of misunderstanding.

The foregoing is obviously true of the "classic" plastic arts. With respect to the specific arts of cyberculture, the virtual remains their natural setting; brick-and-mortar museums can only provide an imperfect projection of those works. We cannot "exhibit" a CD-ROM or a virtual world. We can only navigate them, immerse ourselves in them, interact with them, participate in processes that demand our time. An unexpected consequence of this is that for the arts of the virtual, "originals" are bundles of events in cyberspace; "reproductions" in museums will provide only an impoverished experience of their nature.

The genres of cyberculture are similar to *performance* art, such as dance or theater, the collective improvisations of jazz, the commedia dell'arte, or the traditional poetry competitions of Japan. Like installation art, they demand the active involvement of the receiver, his or her displacement in a symbolic or real space, the conscious participation of the receiver's memory in the construction of the message. Their center of gravity is a subjective process, which frees them from any spatio-temporal closure.

By organizing our participation in events rather than spectacles, the arts of cyberculture will rediscover the great tradition of games and rituals. The most contemporary is joined to the most archaic, to the very origins of art in anthropology; for isn't the very essence of civilization's great ruptures and true "progress"—actively criticizing the tradition with which it breaks—paradoxically a return to the beginning? In games and rituals, neither the author nor the archive is important, just the collective act, here and now.

An engineer of worlds ahead of his time, Leonardo da Vinci organized revels for the princes of Italy, which the crowds animated with their costumes and dances, their ardent existence ... and of which nothing remains. Although we cannot join them, other celebrations are being prepared for tomorrow.

CHAPTER 10

The New Relationship to Knowledge

Education and Cyberculture

Any analysis of the future of education and training in cyberculture must take into account contemporary changes in our relationship to knowledge. What is most apparent in this context is the speed at which knowledge and skills appear and are renewed. For the first time in the history of humanity, the majority of the skills acquired by a person at the beginning of his or her professional career will be obsolete by the end. We also find that the nature of work has changed, with the transaction of knowledge continuing to increase. More and more, work involves learning, transmitting skills, and producing knowledge. Cyberspace also supports intellectual technologies that amplify, externalize, and modify a number of human cognitive functions: memory (databases, hyperdocuments, binary files), imagination (simulation), perception (digital sensors, telepresence, virtual reality), and reasoning (artificial intelligence, modeling complex phenomena). These intellectual technologies promote the following:

- new forms of access to information: navigation of hyperdocuments, the use of search engines to locate information, knowbots or software agents, contextual exploration with dynamic data maps
- new forms of reasoning and understanding such as simulation, an industrialization of thought that is based neither on logical deduction nor on experience-based induction

Because these intellectual technologies, and especially dynamic memory, are objectivized in the digital documents and software available on the network (or easily reproduced and transferred), they can be shared among a large number of individuals, thus increasing the potential for collective intelligence.

Knowledge as flow, work as a transaction of the understanding, and the new technologies of individual and collective intelligence profoundly change the data used for education and training. What must be learned can no longer be planned or precisely defined in advance. Skills and their acquisition are singular and less and less frequently channeled into programs or courses that are valid for everyone. We need to construct new models of the knowledge space. Traditional representations of learning in parallel, graded steps, pyramids structured into levels and organized on the basis of prerequisites, converging toward "higher" forms of knowledge, should be replaced by the image of emerging knowledge spaces, open, continuous, in flux, nonlinear, reorganizing themselves according to the goal or context, where each participant occupies a singular and evolving position.

Two major reforms of education and training are needed at this time. First, the tools and attitudes characteristic of open distance learning must become an integral part of our educational systems. Open distance learning exploits some of the techniques of distance learning, including hypermedia, interactive networks, and all the intellectual technologies of cyberculture. But what is essential is a new style of pedagogy, which promotes both personalized learning and cooperative networked learning. In this context, the teacher inspires the collective intelligence of groups of students rather than directly dispensing knowledge.

Second, the recognition of acquired skills needs to change. Since people learn from their social and professional experiences, since schools and universities are gradually losing the monopoly of creating and transmitting knowledge, public educational systems can at least attempt to help individuals find their way through the knowledge space and foster recognition of the entire range of skills possessed by individuals, including nonacademic skills. The tools of cyberspace enable us to envisage vast systems of automated tests, accessible at any moment, and transactional networks where the supply and demand of skills can be negotiated. The universities of the future, organizing communication among employers, individuals, and apprenticeship resources, will help promote a new economy of knowledge.

I'll develop these ideas further in this and the following chapter and suggest several practical solutions (knowledge trees) to develop learning in the new social context.

A Multitude of Viewpoints

In one of my courses at the University of Paris VIII, entitled "Digital Technologies and Cultural Mutations," I asked each student to prepare a ten-minute report for class. The day before the presentation, the student had to provide me with a two-page summary and bibliography, which could be photocopied by other students interested in the topic.

In 1995 a student handed me the two-page summary with the slightly cryptic remark "Here. It's a virtual report." I looked through the report, which was about digital musical instruments and looked much like all the others: a boldface title, subtitles, underlined words in a fairly well structured text, a bibliography. Amused by my skepticism, he led me to the computer room, and followed by a few other students, we gathered around the screen. I then discovered that the two-page summary I had examined on paper was a print version of a series of Web pages.

Instead of a localized text, permanently fixed to a cellulose medium, instead of a small territory and a proprietary author, a beginning and end, margins forming borders around a text, I was confronted with a dynamic, open-ended, ubiquitous document that linked me to a practically infinite corpus. We speak of "pages" in both cases, but the first page is a *pagus*, a bound field, appropriated, scattered with enracinated signs, whereas the other is a unit of flux, subject to the bandwidth limitations of the network. Even when it refers to other articles or books, the first page is physically closed. The second, however, technically and immediately connects us to the pages of other documents, scattered across the planet, which also refer, indefinitely, to other pages, to other drops in the same global ocean of fluctuating signs.

Beginning as the invention of a small team at CERN, the World Wide Web has spread among Internet users like wildfire, to become one of the principal axes of the development of cyberspace. Obviously this is no temporary trend. I want to posit that the irresistible growth of the Web is indicative of some essential traits of a culture struggling to be born. Bear this in mind as I continue with my analysis.

A Web page is an element, a part of the ever elusive corpus of documents on the Web. But through the links it establishes with the rest of the network, through the crossroads and turnoffs it offers, it also constitutes an organizing selection, a structuring agent, a filter for this corpus. Every element of this uncircumscribable skein is both a packet of information and an instrument of navigation, an item of inventory and original point of view of that inventory. On one side, the Web page is a drop from a leaky faucet; on the other, it serves as a singular filter of the ocean of information.

On the Web, everything exists on the same plane. Yet everything is distinct. There is no absolute hierarchy; each site is an agent of selection, linkage, or partial organization. Far from being an amorphous mass, the Web articulates an open multitude of points of view. But this articulation operates nonhierarchically, rhizomatically, without any god-like point of view, any overarching unification. That this state of things engenders confusion is common knowledge. New search and retrieval tools must be devised, as shown by the amount of current work being done on the dynamic mapping of data spaces, intelligent agents, and the cooperative filtering of information.

The Second Deluge and the Inaccessibility of Everything

Lacking any semantic or structural closure, the Web is not temporally static. It is continuously expanding, moving, and transforming itself. The World Wide Web is a flux. Its innumerable sources, its turbulence, its irresistible swell provide us with a striking image of the contemporary flood of information. Every memory store, every group, every individual, every object can become a transmitter and swell the flood. Roy Ascott speaks, imagistically, of a *second flood*, the flood of information. For better or worse, this flood will never subside. We will have to learn to accommodate its profusion and disorder. Aside from some cultural catastrophe, no sense of order, no central authority will be able to lead us back to terra firma or the stable and well-delineated landscapes that existed before the flood.

The historical pivot of our relationship to knowledge no doubt occurred at the end of the eighteenth century, at that moment of fragile equilibrium when the Old World gave off its most brilliant sparks while the smoke of the industrial revolution began to change the color of the sky. It was the time when Diderot and d'Alembert published their *Encyclopedia*. Until then a small group of men might hope to master all knowledge (or at least its principal elements) and provide others with the ideal of this mastery. Knowledge was still totalizable, aggregate. Starting in the nineteenth century, with the enlargement of the world, the gradual discovery of its diversity, and the increasingly rapid growth of scientific and technical knowledge, the hope of an individual's or small group's mastering knowledge became increasingly illusory. Today it is obvious, tangible, that knowledge has finally become untotalizable, uncontrollable.

The emergence of cyberspace does not at all mean that the "whole" is finally accessible, but rather that the whole is finally out of reach. What should we save from the flood? To assume that we can construct an ark that would contain only the most "important" items would be to yield to the illusion of totality. All of us—institutions, communities, human groups, individuals—need to construct meaning, organize zones of familiarity, tame the ambient chaos. Although each of us must re-construct partial totalities of our own, using individual criteria of per-tinence, these zones of appropriate meaning must also be mobile, changing, in the process of becoming. In place of the image of the great ark, we need to substitute that of a fleet of small arks, boats, and sampans, a myriad of small totalities, different, open, and provisional, secreted by active filtering, perpetually updated by intelligent collectives that crisscross, hail, collide, and mix with one another on the great sea of the informational deluge.

The metaphors currently most central to the relationship to knowledge are navigation and surfing, which imply an ability to confront waves, eddies, currents, and contrary winds across a vast, borderless and ever changing plane. The old metaphors of the pyramid (the pyramid of knowledge), the ladder, and the course (delineated in advance for us) are fragrant with the rigid hierarchies of the past.

The Reincarnation of Knowledge

Web pages express the ideas, desires, knowledge, and offers of transaction of persons and groups. Behind the great hypertext teems the multitude and its relationships. In cyberspace knowledge can no longer be conceived as something abstract or transcendent. It becomes increasingly visible—and tangible in real time—the more it expresses a population. Web pages are not only signed like paper but often lead to direct, interactive communication through e-mail, message boards, or other forms of communication in virtual worlds, such as MUDs and MOOs. Contrary to what the media vulgate would have us believe regarding the "coldness" of cyberspace, interactive digital networks are powerful factors in personalizing and embodying understanding.

It is worth remembering the inanity of the theory of substitution. Just as the telephone didn't prevent people from meeting one another face-to-face, so e-mail communication is often a prelude to physical travel, conferences, and business meetings. Even when no actual meeting takes place, interaction in cyberspace remains a form of communication. Yes, we often hear it said that some people remain hours in front of their screen, isolated from the outside world. Obviously, excess shouldn't be encouraged. But do we say of someone who reads that he "sits for hours on end in front of his books"? No. Because the person who reads has no relationship to the sheet of cellulose; he is in contact with a discourse, a voice, a universe of signification, which he helps construct, inhabits through his reading. The fact that the text is displayed on screen changes nothing. It is still reading, even if, as we have seen, through the use of hyperdocuments and generalized interconnections, the modalities of reading are changing.

Although information media do not automatically determine a given content of knowledge, they do help structure society's "cognitive ecology." We think with and in groups and institutions that tend to reproduce their idiosyncrasies, impregnating us with their emotional climate and cognitive abilities. Our faculties of understanding function with languages, systems of signs, and the intellectual processes furnished by a culture. We don't multiply the same way with knotted strings, stones, Roman numerals, Arabic numerals, abaci, slide rules, or calculators. Cathedral windows and television screens provide very different images of the world, do not incite the same reveries. Certain representations can't survive for long in a society without writing (numbers, tables, lists), whereas we can easily archive them as soon as we have access to artificial memory. To encode knowledge, preliterate societies developed techniques of memory based on rhythm, narrative, identification, physical participation, and collective emotion. But with the rise of writing, knowledge was able to detach itself partially from personal or collective identities, become more "critical," strive for a kind of objectivity and universal theoretical scope. It is no longer simply our modes of knowledge that depend on information media and communications technologies but, through the cognitive ecologies they condition, society's values and judgment criteria as well. Yet it is precisely our criteria for evaluating knowledge (in the broadest sense of the term) that are implicated in the growth of cyberculture, with the likely—and already observable—decline of the values prevalent in civilizations structured by static writing. It is not so much that these values are destined to disappear but rather that they are going to become secondary and lose their power of control.

Perhaps more important than the genres of knowledge and the value criteria that polarize them, every cognitive ecology promotes certain agents whom it places at the center of the processes for accumulating and exploiting knowledge. Here the question is no longer how, or according to what criteria, but who.

In preliterate societies, practical, mythic, and ritual knowledge is embodied by the living community. Every time an old person dies, a library goes up in flames. With the coming of writing, knowledge is embodied in the book. The book is unique, indefinitely interpretable, transcendent, all-encompassing: the Bible, the Koran, sacred texts, the classics, Confucius, Aristotle ... with books, it is the interpreter who masters knowledge.

With the invention of printing, a third type of knowledge has come into being, haunted by the figure of the scholar or scientist. Here knowledge is embodied no longer in the book but in the library. The *Encyclopedia* of Diderot and d'Alembert is less a book than a library. Knowledge is structured through a network of cross-references, possibly already inhabited by hypertext. Here the concept, abstraction, and system serve to condense memory and ensure the intellectual control that the inflation of knowledge already endangers.

The deterritorialization of the library that is occurring today may be only the prelude to the appearance of a fourth type of relationship to knowledge. Through a kind of spiraling return to the orality of our origins, knowledge may again be transmitted through living human collectives rather than distinct media served by interpreters or scholars. Only this time, unlike archaic orality, the direct transmitter of knowledge will no longer be the physical community and its carnal memory but cyberspace, the region of virtual worlds, through which communities discover and construct their objects and recognize themselves as intelligent collectives.

From now on, abstract systems and concepts will give way to highly granular maps of singularities, the detailed description of great cosmic objects, the phenomena of life or human customs. All the major contemporary technological and scientific projects, particle physics, astrophysics, the human genome, space, nanotechnology, ecological and climatic monitoring—they are all dependent on cyberspace and its tools. Image databases, interactive simulations, and electronic forums provide us with a better understanding of the world than theoretical abstraction, which now fades into the background. These will define the new standard of knowledge. And these tools will help ensure the efficient coordination of the producers of knowledge, where theories and systems tend to give rise to allegiance or conflict.

It is striking to discover that some scientific experiments performed in the large particle accelerators require so many resources, are so complex, so difficult to interpret, that they can take place only once, or nearly so. Every experiment is singular. This seems to contradict the ideal of reproducibility of classical science. However, these experiments are still universal, although we are unable to reproduce them. A multitude of scientists from around the world participate in them, forming a kind of microcosm or projection of the international community. But it is especially direct contact with the experiment that has almost entirely disappeared, replaced by the large-scale reproduction of digital data. These data can be consulted and processed in a number of labs around the world using the instruments of communication and data processing provided by cyberspace. In this way the entire scientific community can participate in these highly specialized experiments, which are also *events*. Universality is based on the interconnection in real time of the scientific community, its global cooperative participation in the events that concern it, rather than depreciation of the singular event that typifies the former universality of the exact sciences.

Simulation

Simulation plays a central role among the new modes of understanding made possible by cyberculture. Simulation is an intellectual technology that enhances individual imagination (augmented intelligence) and enables groups to share, negotiate, and refine shared mental models, regardless of their complexity (augmented collective intelligence). To augment and transform certain human cognitive abilities (memory, imagination, calculation, expert reasoning), computer technology partially externalizes these abilities on digital media. Yet once such cognitive processes have been externalized and reified, they become shareable and reinforce the process of collective intelligence—assuming, of course, that the underlying technologies are used appropriately.

Even expert systems (or knowledge-based systems), which are traditionally classified under the heading "artificial intelligence," should be considered technologies for communication and the rapid mobilization of practical skills in organizations, rather than clones of human experts. In terms of cognition as well as the organization of work, intellectual technologies must be thought of in terms of articulation and synergy rather than substitution.

Simulation technologies, especially those that make use of interactive graphics, do not replace human reasoning but extend and transform the capacities of imagination and thought. Our long-term memory can store an extremely large quantity of information and knowledge. However, our short-term memory, which contains the mental representations we use in conscious thought, is limited. It is, for example, impossible for us to represent clearly and distinctly more than approximately ten interacting objects.

Although we can mentally evoke the image of the chateau of Versailles, we are unable to count all its windows "in our head." The degree of resolution of the mental image is inadequate to the task. To achieve this level of detail, we require the use of an external auxiliary memory (engraving, painting, photograph) with which we can perform new cognitive operations: counting, measuring, comparing, and so on. Simulation is an aid to short-term memory, which works with dynamic complexes rather than static images, text, or tables of numbers. The ability to easily vary the parameters of a model and immediately, and visually, observe the consequences of this change can be considered an enhancement of the imagination.

Today simulation plays an increasingly important role in scientific research, industrial design, management, and learning, as well as games and entertainment (especially interactive games). Neither theory nor experiment but a way of industrializing the experience of thought, simulation is a special mode of awareness unique to a developing cyberculture. The principal goal of research is not to replace experience or serve as a substitute for reality but to enable us to rapidly formulate and explore a large number of hypotheses. From the point of view of collective intelligence, it enables us to visualize and share virtual worlds and universes of signification of great complexity.

Knowledge is now encoded in on-line databases, in maps that are updated in real time with global phenomena, and in interactive simulations. The efficiency, heuristic fecundity, power of mutation and bifurcation, and temporal and contextual pertinence of the models have supplanted the older criteria of objectivity and abstract universality. Yet the availability of connectivity, the use of standards and formats, and compatibility and planetary interoperability have given us a more concrete form of universality.

From Chaotic Interconnectivity to Collective Intelligence

Detotalized knowledge fluctuates. The result is a violent sentiment of disorientation. Should we hold on to the procedures and models that gave us the old orders of knowledge? Or should we instead leap forward and jump headfirst into the new culture, which offers us specific remedies for the ills it engenders. Many-to-many real-time interconnectivity is certainly the cause of the disorder. But it's also the necessary condition for practical solutions to the problems of orientation and learning in the universe of knowledge in flux. Interconnectivity promotes the processes of collective intelligence in virtual communities through which the individual is better able to confront informational chaos.

Indeed, the mobilizing ideal of information technology is no longer artificial intelligence (make a machine more intelligent, and possibly more intelligent than humans) but *collective intelligence*; namely, the enhancement, optimal use, and fusion of skill, imagination, and intellectual energy, regardless of their qualitative diversity. This ideal of collective intelligence obviously involves the sharing of memory, imagination, and experience through the widespread exchange of knowledge, new forms of flexible organization and coordination in real time. Although new communications technologies enable human groups to function as intelligent collectives, they do not automatically determine them. The defense of exclusive power, institutional rigidity, and mental and cultural inertia can obviously result in social uses for new technologies that are far less positive on the basis of humanist criteria.

Cyberspace, the interconnection of computers across the planet, is becoming the major infrastructure of production, management, and economic transaction. It will soon constitute the principal international collective infrastructure of memory, thought, and communication. Within a few decades, cyberspace, virtual communication, its imagery and interactive simulations, the irrepressible turbulence of texts and signs, will be the essential mediators of humanity's collective intelligence. This new medium of communication and information will be accompanied by previously unknown forms of understanding, new participants in its production and management, and new criteria of evaluation in the orientation of knowledge. Any educational policy worth the name will have to take these factors into consideration.

CHAPTER 11

Education and the Economy of Knowledge

Open Distance Learning

Today's educational systems are subject to new constraints in terms of quantity, diversity, and the rate of change of knowledge. From a purely quantitative point of view, the need for training has never been as great as it is today. In many countries, a *majority* of a given age group are enrolled in secondary education programs. Universities are filled to capacity. Facilities for continuing and professional training are turning away students. Nearly half of society is, or would like to be, in school.

We will be unable to increase the number of teachers in proportion to the demand for training, which has become increasingly diverse and widespread in countries around the world. In the poorer countries, the cost of education is a major concern. We need to find solutions based on technologies that can multiply the educational efforts of teachers and trainers. Audiovisual, interactive multimedia, computer-based training, educational television, cable, traditional techniques of distance training based on writing, telephone, fax, and Internet tutoring—all of these technological solutions, of varying applicability depending on their content, the situation, and the needs of the learner, have potential utility and have been abundantly tested. In terms of both material infrastructure and the cost of operation, "virtual" schools and universities are less expensive than traditional brick-and-mortar institutions offering on-site training.

The demand for training has not only experienced enormous quantitative growth but also undergone a profound qualitative mutation that is felt as a growing need for diversification and personalization. Individuals are increasingly less willing to follow a rigid series of courses that do not correspond to their real needs and the specificity of their experiences. Responding to this increased demand by simply increasing supply would be an outdated, "industrial" response, poorly adapted to the flexibility and diversity now required of our educational systems.

We have seen how the new paradigm of navigation (antithetical to the rigid programming of traditional education) associated with the retrieval of information and cooperative learning in cyberspace can be used as a guide to widespread, personalized access to knowledge. Universities and, increasingly, primary and secondary schools are providing students with opportunities for navigating the ocean of information and knowledge accessible through the Internet. Educational programs can be monitored remotely on the World Wide Web. E-mail and newsgroups can be used for intelligent tutoring and cooperative apprenticeship programs. Hypermedia (CDs, interactive on-line multimedia databases) provide rapid, intuitive access to vast collections of information. Simulation programs can help learners familiarize themselves with complex phenomena at low cost without being subject to situations that are dangerous or difficult to control.

Education specialists acknowledge that the distinction between on-site and distance learning will become increasingly less relevant as the use of telecommunications networks and interactive multimedia are gradually incorporated into traditional methods of education.[1] Distance learning has for a number of years been viewed as something nonessential by the educational community; it will soon become, if not the norm, at least a trailblazer in educational reform. The characteristics of distance learning are similar to those of the information society as a whole (networking, speed, personalization, etc.). Moreover, this type of education is in sync with the "learning organization," which a new generation of managers are trying to establish within the enterprise.

Cooperative Learning and the New Role of the Teacher

The key point here is the qualitative change in the learning process. It is less important to convert conventional programs into interactive hypermedia formats or "abolish distance" than it is to implement new paradigms for acquiring knowledge and building skills. The most promising direction, which reflects the outlook of collective intelligence in the educational field, is that of cooperative learning.

Some computerized systems for group learning are specially designed for sharing databases and incorporating e-mail and virtual meetings. Such systems are known as computer-supported cooperative learning (CSCL). On the new virtual campus, professors and students will share the available material and informational resources. Teachers will learn along with their students and continuously update their knowledge along with their teaching skills. (The continuous training of teachers is one of the most obvious applications for open and distance learning.)

The most recent information will be easily and directly available through on line databases and the World Wide Web. Students will be able to participate in deterritorialized virtual conferences, where the best researchers in the field will be present. The primary role of education will no longer be the distribution of knowledge that can now be obtained more efficiently by other means. It will help provoke learning and thinking. Education will become a driving force of the collective intelligence for which it is responsible. It will focus on managing and monitoring learning: encouraging people to exchange knowledge, relational and symbolic mediation, personalized guidance for apprenticeship programs, and so forth.

Public Regulation of the Knowledge Economy

There are now a range of attitudes and approaches regarding the use of new technologies in education. For example, a number of studies have been conducted on the use of multimedia as a distribution medium or of computers as a tireless substitute for teachers (computer-assisted instruction). In this scheme of things—as conventional as any—information technology supplies teaching machines. Other approaches treat computers as instruments of communication, information retrieval, calculation, and message production (text, image, sound) placed in the hands of "learners."

The perspective adopted here is somewhat different. The increasing use of digital technologies and interactive communications networks has engendered a profound mutation in our relationship to knowledge, which I attempted to outline in the previous chapter. By extending certain human cognitive capacities (memory, imagination, perception), digitally based intellectual technologies redefine their scope, significance, and sometimes even their nature. The new possibilities for distributed collective creation, cooperative learning, and networked collaboration provided by cyberspace challenge the way our institutions operate and conventional methods for the division of labor in business as well as education.

How will we ensure that educational practices remain in step with these new processes for the transaction of knowledge? It's not a question of using technology at any cost but a question of consciously and deliberately accommodating a change in civilization that seriously questions the institutional forms, attitudes, and culture of traditional education systems, and especially the role played by teacher and student.

Cyberculture's great promise, in terms of lower costs as well as access to education, is not the transition from on-site to distance learning or from traditional literate and oral traditions to multimedia. It is the transition from highly institutionalized forms of education and training (schools, universities) to a generalized situation for the exchange of knowledge, the creation of an autodidactic society, the self-managed, mobile, and contextual recognition of skills. In this context, the role of government should be to

- guarantee everyone some form of high-quality elementary education;[2]
- provide everyone with unrestricted, free access to media libraries,

orientation, documentation, and self-training centers, and points of entry into cyberspace, without neglecting the *human mediation* so essential in accessing knowledge;

- regulate and promote a new economy of knowledge in which each individual, each group, each organization is considered a potential learning resource in a continuous and personalized system of training.

The Stream of Knowledge

Ever since the end of the sixties, we have been experimenting with a way of relating to knowledge and skills that was unknown to our ancestors. In the past, the skills acquired during our youth were generally still in use at the end of active life. These skills were transmitted nearly unchanged to young workers or apprentices. Of course, new procedures and new techniques were introduced. But *on the scale of a human life*, the bulk of a person's useful skills were durable. Today the majority of the skills acquired at the beginning of a career are obsolete at the end of professional life, or sometimes before. Economic disturbances, like the rapid rate of change in science and technology, determine the general acceleration of social time. Consequently individuals and groups are confronted no longer with stable skill sets, forms of knowledge that are passed on and reinforced by tradition, but with a chaotic stream of knowledge whose course is difficult to predict. The intense relationship to apprenticeship, to the transmission and production of knowledge, is no longer reserved for an elite but now affects the mass of people in their daily life and work.

The old pattern in which the skills learned in our youth would serve us for the rest of our life no longer holds. Individuals are expected to change professions several times during their career, and the very notion of a specific craft is becoming increasingly problematic. It would be better to say that each of us has a singular collection of varied skills. We would therefore be responsible for maintaining and enriching our skill set throughout our lifetime. This approach calls into question the classic division between a period of apprenticeship and working career (since we now learn all the time) as well as the idea of a job or profession as the principal means of a person's economic and social identification.

Through continuing education, sandwich training, in-house learning, and participation in community groups and unions, a continuum is being formed between training, on the one hand, and professional and social experience, on the other. Within this continuum, all forms of skills acquisition (including self-teaching) will play a legitimate role.

For a growing proportion of the population, work is no longer the repetitive execution of a predefined task but a complex activity in which the inventive resolution of problems, teamwork, and human relations management play nonnegligible roles. The exchange of information and knowledge (production of knowledge, learning, transmission) is an integral part of professional activity. Through the use of hypermedia, simulation, and cooperative learning networks in the workplace, professional training in the corporate sector is becoming increasingly integrated with production.

The older relationship to skill was substantial and territorial. Individuals were recognized by their degrees, which were associated with a specific discipline. Employees were identified with their jobs, which typified a trade and fulfilled certain functions. In the future, we will be much more involved in managing processes: trajectories and cooperation. The various skills acquired by individuals based on singular life experiences will help construct collective memories. Accessible on-line, these digitally based dynamic memories will serve concrete needs, here and now, of individuals and groups in a work or learning environment (which will be indistinct). In this way, the virtualization of "networked" organizations and companies will soon correspond to a virtualization of the relationship to knowledge.

The Recognition of Acquired Skills

Obviously education must prepare itself for this new universe of work. However, we must also acknowledge the educational or formative character of a number of economic and social activities, which raises the question of their recognition or official validation, since the current degree system is beginning to appear less and less adequate. Moreover, the time needed to institute new degree programs and design courses is no longer in phase with the rate of change of knowledge.

It may seem banal to acknowledge that all forms of education and training must be capable of resulting in a certification or socially recognized validation of some kind. Yet we are currently far from this goal. A large number of the apprenticeship processes that take place within formal structures for continuing education, not to mention the skills acquired during an individual's social or professional life, result in no official recognition. Our relation to emerging knowledge casts doubt on the traditional interdependence of two functions of the educational system: teaching and the recognition of skills.

Because individuals are increasingly going outside the academic network for training, it is up to our educational systems to organize procedures for recognizing the knowledge and skills acquired during social and professional life. In this sense, public services that use multimedia technologies (automated testing, simulators) and interactive networks (computerized testing and skill recognition, on-line exam tutors) on a large scale could relieve educators and traditional educational institutions of a responsibility for monitoring and approval that is less "distinguished" than, but just as necessary as, traditional apprenticeship programs. Through such an extensive, decentralized, and open system for the recognition and approval of skills, the most diverse methods of learning, even the least formal, could be sanctioned through some form of individual certification.

The evolution of the educational system can't be dissociated from the system for the recognition of skills that accompanies and drives it. For example, we are well aware that exams are largely responsible for structuring teaching programs. For this reason, incorporating new technologies for education and training without changing the mechanisms used to validate training will only build up the body of the educational institution while blocking the development of its senses and intellect.

A controlled deregulation of the current system for recognizing skills could promote the development of sandwich training and all forms of training that place significant emphasis on professional experience. By promoting the development of original modes of validation, this deregulation would also encourage education through collective exploration, along with other initiatives that fall halfway between social experimentation and explicit training. Such a development could not fail to produce interesting results with certain forms of educational training, which are frequently hampered by teaching styles that are unlikely to mobilize initiative and strictly oriented toward the final granting of a degree.

Within a broader context, the controlled deregulation of skill recognition would stimulate socialization of the traditional functions of the school. It would make use of all the available resources in providing personalized learning trajectories, adapted to the varied goals and needs of individuals and the communities involved.

The industrial and commercial performance of companies, regions, and large geopolitical zones is tightly correlated with knowledge management policies. Knowledge and skills are today the principal sources of wealth for companies, cities, and nations. Yet we are having significant difficulty managing these skills, at both the community and regional level. On the demand side, we find that there is an increasing lack of correlation between available skills and economic demand. On the supply side, a large number of skills are neither recognized nor identified, especially among those without diplomas. These phenomena are particularly acute where industrial reconversion is taking place or where there is a delay in the development of entire regions. Along with rethinking our attitude toward diplomas, we must also conceive of methods of skill recognition that are capable of high visibility within a network of skills and can retroactively use demand dynamically to drive supply. Once we accept the fact that the acquisition of skills must result in explicit social recognition, the problems of skills management, at both the corporate and local level, will be, if not fully resolved, at least attenuated. In the following chapter, I'll look at an example of skills management that implements the foregoing approach by using new tools for interactive communication.

CHAPTER 12

The Knowledge Tree

Continuous personalized training, the orientation of learners inside a fluctuating and detotalized knowledge space, cooperative apprenticeship, collective intelligence in virtual communities, partial deregulation of the modes of recognizing knowledge, the dynamic management of skills in real time—social processes such as these actualize the new relationship to knowledge. Michel Authier and I have designed a computerized system running on a network that can coexist with, integrate, and synergize these different processes in a positive fashion. The *knowledge tree* is a digital tool for overall management of skills in educational institutions, companies, job pools, local communities, and associations.[1] It is currently being used on an experimental basis throughout Europe, and especially in France, where the software has been installed at Électricité de France and PSA (Peugeot and Citroën), midsize companies, universities, business schools, local communities, and subsidized housing developments.

Using this approach, each member of a community can spread awareness of the diversity of his or her skills, even those that are not approved by any traditional university or educational system. Using descriptions provided by individuals themselves, a knowledge tree makes visible the organized multiplicity of skills available within a community. It is a kind of dynamic map, viewable on-screen and resembling a tree in appearance, where each community generates a tree of a different shape.

Skills include behavioral abilities *(knowledge-of-being)* as well as know-how and theoretical knowledge. Each elementary skill possessed by an individual is recognized by means of a *brevet*, based on a precisely specified procedure (testing, nomination by one's peers, the supply of proof of some sort, etc.). Visible on-screen, the dynamic map of a group's know-how doesn't result in any a priori classification of skills: *automatically* produced by software, it is the real-time expression of the trajectory of apprenticeship and experience of members of the community. The tree of a community grows and is transformed as that community's skills evolve. Thus brevets for basic forms of knowledge will be located in the trunk. Brevets for highly specialized forms of knowledge, associated with more extensive education, will form the leaves. The branches will combine skills that are almost always associated with one another in the individual lists of individual skills. But the organization of knowledge expressed by a tree is not permanent. It reflects the collective experience of a human group and thus evolves with that experience. For example, a brevet located on a leaf at time *t* can be found lower down on a branch at time *t* + *n*. The tree, which is different for each community, doesn't reflect the customary partitions into disciplines, levels, successive phases of study, or institutional divisions. On the contrary, the mechanism for dynamic indexing and navigation that it offers produces a knowledge space without separations, one that is continuously reorganized depending on context and use.

Representation as a knowledge tree means that simple inspection can be used to locate the position occupied by a given kind of knowledge at a given moment, and the possible apprenticeship itineraries for achieving a given skill. Each individual has a personal image (an original distribution of brevets) on the tree, an image that he or she can refer to at any moment. This image is called the person's *blazon* to indicate that today's true nobility has been conferred on the basis of skill. Thus people obtain a better awareness of their situation in the "knowledge space" of the communities in which they participate and can consciously elaborate their own learning strategies.

Electronic messages, "routed on the basis of knowledge," correlate the supply and demand of know-how within the community and announce the availability of training and exchange for each elementary skill. Such an instrument serves the social bond through the exchange of knowledge and the employment of skills. All transactions and queries recorded by the software help to continuously determine the value (always contextual) of elementary skills as a function of different economic, educational, and social criteria. This evaluation, extended through use, is an essential mechanism of self-regulation.

Within a community, the system of skill trees can contribute to the struggle against exclusion and unemployment by recognizing the abilities of individuals without diplomas, by favoring a better adaptation of training to employment, and by stimulating a genuine "skills market." Within the network of schools and universities, the system can be used to implement a decompartmentalized and personalized cooperative pedagogy. Within an organization, knowledge trees provide instruments for identifying and mobilizing abilities and evaluating training, as well as a strategic vision of the evolution of, and need for, skills.

Because all types of learning methods can result in some form of qualification, the "tree" provides for improved skills management. Similarly, by evaluating the signs of a person's competence in real time, skills management helps validate those qualifications. By assigning themselves a certain number of signs of competence, individuals become simultaneously accessible on the network. The individual is indexed to the navigation space and can be contacted for mutual exchanges of knowledge or skill requests. An improvement of the process of qualification thus has positive effects on sociability. This instrument makes visible in real time the rapid evolution of highly diverse skills. By allowing the full range of skills to be expressed, it doesn't compartmentalize individuals within a job or category and promotes continuing personal development.

Today every country has a diploma system and a structure for recognizing different types of knowledge. Moreover, within a given country, diplomas—notoriously lacking in this respect—are the only system of representation for skills shared by all branches of industry, all business and social environments. For the rest of society, extreme heterogeneity is commonplace. The knowledge tree, however, can translate other systems of knowledge recognition and *pool* the signs of competence.

Nectar: The Knowledge Tree Enacted

During 1994 and 1995, the business administration departments of five universities conducted an international knowledge tree project, sponsored by the European Union: the University of Aarhus in Denmark, the University of Sienna in Italy, the University of Limerick in Ireland, the University of Lancaster in England, and the University of Geneva in Switzerland. The project, known as Nectar (Negotiating European Credit Transfer and Recognition), was intended to facilitate student exchange in Europe through the cooperative construction of a common system of knowledge recognition. Right now it is difficult to establish equivalences between European diplomas, and even more difficult when trying to compare different years, semesters, or course modules. The procedure was as follows: A number of brevets were assigned to each course given by university departments in five countries. These brevets corresponded to skills normally acquired by students who had successfully completed the course work. The translation of the course credits into brevets, suggested by the project's international team, was approved, and sometimes modified, by all the teachers involved. In this way a transition was made between a formal educational approach and one based on acquired skills.

One difficulty arose from the fact that program boundaries, course titles, and contents were different in the participating universities, but the language used to describe students' skills by means of brevets (finer, more "microscopic" than conventional methods) had to be identical. Surprisingly, this objective was achieved without too much difficulty, primarily by using a mailing list connecting all the participants. Once the courses had been translated into brevets, it was easy to create knowledge trees based on the universities' records of student grades. Trees for each of the five universities could be viewed individually, but a large, common tree combined students from the five institutions. In this way each student was able to

- communicate with other students based on their skill profile and the courses they were taking
- monitor his or her position on the common tree
- determine any additional skill profile he or she wanted to acquire
- view a real-time description of all the courses (from the five universities), which enabled him or her to easily compare skills to a target profile

Note that students were first encouraged to examine the skills they wanted to acquire (with the assistance of the tree) and only then look up information about the courses they might take to achieve those skills. In this way students could decide where to study (Sienna, Aarhus, Lancaster, etc.) with a better understanding of their objectives. The mailing list enabled students to obtain firsthand information about courses in another university. The student could then prepare a trip by viewing the position his or her blazon would have in the university he or she would be attending. The students' blazons generally assumed a "normal" position on the tree of their originating university but were "skewed" on the tree of the host university, to which they often brought new skills.

During the project, the international team was able to measure the point at which the concept of prerequisites becomes relative. For example, in Anglo-Saxon universities, brevets indicating skills in theory or history are generally found high on the tree (acquired fairly late in the program), whereas practical skills and case studies were found on the trunk (acquired at the start of the program). The respective positions for brevets corresponding to these skills were reversed on the trees of universities that were part of the "Latin" tradition.

One of the principal advantages of the approach validated by the Nectar project is obviously international interaction and the optimization of university resources. Although the same language (brevets indicate skills) was used everywhere, the specificity of each cultural and institutional environment was respected, since each university gave rise to a different tree, reflecting the originality of its organization of knowledge. Movement was encouraged. Each student was able to move from one community to another, one country to another, always retaining the same list of brevets that defined his or her skills. This list (possibly enhanced by experience) automatically assumed a different appearance and values on different trees.

A Universal System without Totality

Clearly if such a project could be realized on an international level, in five different countries, it could be conducted in one country, or within a single university, with the same degree of global visibility, resource optimization, decompartmentalization, nonhierarchical cooperation, and mobility. It would also be possible to add to students' blazons the skills from their social or professional experience, to connect employers to university trees, and so forth. The business schools, research organizations, and graduate schools that decided to implement knowledge trees understood this.

The advantages of such a system in an international environment are obvious. It's not a question of normalizing or controlling degrees, since in each particular community, the same brevets, the same profiles (a group of brevets) can assume variable positions and values, corresponding to the characteristics of use and local culture. And yet every student can move from one community to another, one country to another, always retaining the same list of brevets used to define his or her skills. This list automatically assumes a different appearance and different values depending on the tree.

Trees can be combined, split, and connected; small trees can be merged into larger ones. The skill space I am proposing here can gradually be generalized, through extension and connection, without ever imposing any a priori standards. Like the image of an emergent cyberculture, the knowledge tree offers a universal approach (the same system can be virtually used everywhere, enables every form of coordination, transfer, transition, and trajectory imaginable), *but without totalization.*

This is because the nature, organization, and value of knowledge are not fixed but remain in the hands of the various individual communities.

The perspective outlined here doesn't require any centralized decision making or large-scale organization. A local project that focuses on the struggle against exclusion and socialization through apprenticeship can be developed in one location; other projects focused on new methods of training and qualification can be implemented elsewhere. Some might involve a specific initiative for the dynamic indexing of training resources. In a given company or community, new experiments involving new forms of skill management could be organized. Gradual convergence—always voluntary and involving the participation of motivated individuals—is ensured of long-term success through the coherence of the model and its pertinence to the figures emerging from the relationship to knowledge.

CHAPTER. 13

Cyberspace, the City, and Electronic Democracy

For the past few years, urbanists, architects, and generally anyone concerned about management and leadership in local communities have been confronted with an unusual problem: the need to account for new systems of interactive on-line communication. How will the development of cyberspace affect urban life and the development of territorial infrastructures? What positive, active steps can be taken, and what kinds of projects can be implemented that will best exploit the new instruments of communication? These problems are not only the concern of politicians, urbanists, and planners; in fact, their greatest impact is on the common citizen.

Cybercities and Electronic Democracy

Will the development of cyberspace result in the decentralization of the great urban centers, in new forms of distribution for economic activities? It should be pointed out that the current movement to constitute and strengthen the world's giant metropolises has little chance of undergoing any kind of permanent reversal.[1] Statistics show that the greatest concentration of access nodes to cyberspace and the greatest use of digital technologies coincide with the principal global centers of scientific research, economic activity, and financial transactions. The spontaneous effect of the growth of cyberspace is to augment the capacities for strategic control of traditional centers of power by means of increasingly vast and dispersed technological, economic, and human networks.

Nonetheless an affirmative policy on the part of governments, local communities, citizens associations, and groups of entrepreneurs can use cyberspace for the development of disadvantaged areas by fully exploiting its potential for collective intelligence: emphasizing local skills, promoting synergies between resources and projects, exchanging knowledge and experience, setting up mutual assistance networks, promoting increased participation in political decisions, making various forms of expertise and partnership universally accessible. I again want to emphasize that this use of cyberspace does not automatically follow from the presence of a hardware infrastructure but also requires a profound change in attitudes, modes of organization, and political morals.

Rather than become polarized over the use of telecommuting and the substitution of telecommunications for transportation, a new orientation of land use planning policies in the large cities might turn to cyberspace to encourage reconstruction of the social bond, debureaucratize government administrations, optimize a city's resources and infrastructure in real time, and experiment with new aspects of democracy.

Concerning this last point—often subject to misunderstanding—I want to emphasize that the distribution of government propaganda on the network, the identification of the e-mail addresses of political leaders, or the organization of referendums over the Internet are only caricatures of electronic democracy. True electronic democracy consists in using the possibilities for interactive and collective communication offered by cyberspace to encourage the expression and elaboration of urban problems by local citizens themselves, the self-organization of local communities, the participation in deliberations by those directly affected by them, the transparency of public policies and their evaluation by citizens.

Concerning the relation between city and cyberspace, several attitudes have already been adopted by theoreticians and practitioners. They can be grouped into four broad categories:

- the identification of *analogies* between territorial communities and virtual communities

- the *substitution* or replacement of functions provided by the traditional city with the technical services and resources of cyberspace
- the *assimilation* of cyberspace to some kind of urban hardware or conventional territory
- the exploration of different ways of *articulating* the difference between the ways cities operate and new forms of collective intelligence being developed in cyberspace

I want to offer my criticism of the first three attitudes and try to demonstrate why I believe the fourth, the exploration of articulation, is the richest approach for the future.

Analogy, or the Digital City

Can virtual communities be designed based on the existing urban model? One of the best examples of this is the "digital city" in Amsterdam, a free Internet-based service available in Dutch. This digital city duplicates some of the infrastructure and institutions typical of the classic city: administrative information, business hours for municipal services, library catalogs, and so forth. Groups of residents are also allowed to occupy "real estate" in this digital city. They can distribute information and organize electronic conferences. There have been a few highly original discussion groups and electronic newspapers in this digital city, where local politics obviously play a role. Amsterdam's digital city can also access all other existing Internet services: the World Wide Web, e-mail, international discussion groups, and others. Since it was opened the Amsterdam digital city has had uninterrupted growth and remarkable popular success, no doubt largely associated with the fact that communication within it is free (except for the phone call), in Dutch (rather than English), and unrestricted.

Dozens, perhaps hundreds, of cities or regions throughout the world will no doubt conduct similar experiments. The Amsterdam virtual city serves as an important example, and it is in this sense that I would like to talk about it. One of its primary purposes is to sensitize business and political leaders to the new possibilities offered by large-scale digital communication. The key here is "unrestricted access," implying, as it does, the struggle against exclusion and compensation for the disequilibrium between society's information rich and information poor. My purpose is not to condemn this type of experiment and its assumptions, yet I can't help but feel a certain unease with the project's systematic duplication of the institutional terrain in a virtual landscape, which we are seeing pretty much all over.

For example, virtual museums are frequently nothing more than bad catalogs posted on the Internet. However, the very notion of the museum as an "archive" for "conservation" is called into question by the development of a cyberspace in which everything circulates with increasing fluidity and where the distinction between an original and a copy has been blurred. Instead of reproducing conventional exhibits on Web sites or interactive kiosks, why not attempt to design trajectories that can be personalized or constantly redefined through collective navigation in spaces that are completely independent of any physical collection of objects? It would be even better if we were to encourage the creation of new types of work: virtual spaces that could be invested and actualized by their explorers.

Similarly, conventional magazines and newspapers are now online, where they provide a bit more information than the paper version, automatic indexing, and chat groups, which serve as a kind of improved "letters to the editor" section. However, the very structure of media-based communication—a centralized group of transmitters and a public of passive, dispersed receivers—needs to be reevaluated in cyberspace. If an individual can transmit to many others, participate in news and discussion groups with experts, and filter the informational deluge using his or her own criteria (which is becoming technologically feasible), is there any compelling need to rely on the limitations of conventional journalism for our news?

The duplication of conventional institutional forms in cyberspace and "universal access" cannot replace a general policy governing relations between cyberspace and national territories. Even if experiments such as Amsterdam's digital city are essential, they are only a transitional stage in the reassessment of the traditional institutional forms of municipal administration, local newspapers, museums, schools. In every specific situation, the instruments of cyberspace enable us to move toward forms that attenuate the separation between administrators and administered, teachers and students, museum directors and visitors, authors and readers. These new forms of cooperative organization, which are currently being explored in a number of local or international venues in cyberspace, are characterized by the fact that they *maximize* and *pool* an intelligence distributed throughout connected communities and create real-time synergy.

Substitution

The idea of substitution has been promoted largely by "urban and physical planners." The argument is simple. New tools for cooperative labor enable users to participate in the international economy from home or an outside office. Now, for a large number of activities, traveling to a job is no longer a requirement. There are a number of benefits to this arrangement: less crowding in urban centers, improvement of highway congestion, less pollution, better distribution of the population throughout the country, the possibility of a revival of regions affected by desertification and mass unemployment, improved quality of life. A simple economic calculation shows that the overall social cost of teleconferencing is less than that of business trips, that a home computer is less expensive than a few square feet of office space in a city, and so on.

The same approach used for labor can also be applied, in almost identical terms, to higher education and professional training. Why construct brick-and-mortar universities when we can encourage the development of distance learning and interactive, cooperative apprenticeship systems that are accessible throughout the country?

Although I don't want to criticize well-meaning attempts to replace transportation and physical presence with telepresence and interactive telecommunication, I would like to draw attention to the following facts. Simple examination of the data shows that the development of telecommunications parallels that of physical transport: the relationship between the two is direct rather than inverse. In other words, the more we communicate, the more we move around. There are obviously numerous exceptions, but they occur within a dynamic of global growth for interactions and relationships of all kinds. In fact, we're traveling more and more, with the average length of time spent away from home increasing.

The largest class of contemporary distance workers are sales personnel, senior management, and scientists and researchers, who, through the use of on-line services, communications terminals, and remote processing, travel more than in the past while remaining in constant contact with their offices, laboratories, customers, and employers.

As for living and working within a country, note that the increasing delocalization of economic activities also parallels an international movement of increased migration, whether this arises from economic or political reasons, or even warfare. The growth of migration affects scientists as much as so-called unskilled workers. The mobility characteristic of economic activities and populations is affected by the same consequential historical trend toward deterritorialization: they are not mutually substitutable.

With respect to our expectations for "urban and regional planning" based on telecommuting and distance learning, any incidence of delocalization resulting from the increasing use of cyberspace remains ambivalent. It's possible, of course, that delocalization will take place, benefiting areas of Europe that have been affected by deindustrialization or rural exodus. But it may also accelerate desertification of those regions, promoting development in recently created countries where labor costs are lower and social legislation is not as restrictive. This is why data entry or programming in Northern countries is often done by Asian "telecommuters." Moreover, distance learning companies and on-line universities and training institutions are now going after an international market. Frequently originating in the Northern Hemisphere, these companies are beginning to short-circuit national or regional educational systems—with the obvious economic and cultural implications. Rather than reestablish equilibrium among geographic areas, the growing use of cyberspace may further accentuate regional disparities.

Cyberspace is a powerful factor in deconcentration and delocalization, but it doesn't eliminate "centers." Spontaneously, its principal effect would be to render intermediaries obsolete and provide nodes of power with increased capacity for direct control and mobilization of resources, skills, and markets, wherever they happen to be.

My feeling is that a genuine reequilibrium among global regions will be achieved only through the voluntary encouragement of regional initiatives and dynamics that are simultaneously endogenous and outward looking. Once again, the necessary condition is the enhancement and exchange of local skills, resources, and projects, rather than their unilateral subjection to the criteria, needs, and strategies of the dominant geopolitical and geoeconomic centers. Urban and regional planning implies a strengthening of the social bond and collective intelligence. Interactive communications networks are merely the tools for such a policy. The instruments of cyberspace, although they quite naturally reinforce the power of the "centers" by making them ubiquitous, can also support highly granular strategies for forming regional groups into self-organized agents. Computerized methods for mutual listening, resource identification, and real-time cooperation and decision evaluation can serve as a powerful force in strengthening the mechanisms of democracy and economic initiative in underprivileged regions.

With respect to contemporary urban problems, it is unlikely that they will be resolved by well-meaning projects for rural revival or the more or less authoritarian assignment of populations to rural areas crisscrossed by telecommuting hardware. Policies on low-income housing, improving the transportation infrastructure, reducing automobile traffic, promoting electric vehicles, or struggling against social inequality or the poverty of the ghetto will still be necessary, independently of any appeal to the instruments of cyberspace. Obviously I am not opposed to the idea of telecommuting, even if it develops without official encouragement. But in this context, interactive communications networks should have as their primary focus the reconstruction of urban sociability, self-management of the city by its residents, and real-time control of a collective infrastructure. They should not serve as a *substitute* for the concentrated diversity, physical interaction, and direct human exchange that are the principal attractions of urban life.

Assimilation: A Critique of the Information Highway

The third way of looking at the relationship between cyberspace and the city is the *assimilation* of interactive communications networks to the type of infrastructure that is already organizing and "urbanizing" the landscape: railroads, highways, waterways, pipelines, electrical grids, and cable television and telephone networks. Such assimilation, which obviously reinforces certain obvious interests, is the product of a political and administrative technocracy working with the leaders and "communicators" of the large industrial corporations involved. From this point of view, the "information highway" or "multimedia" essentially represent a new hardware, "content," and services *market*, which the telephone, cable, television, publishing, and computer industries fight for tooth and nail. Newspapers are making a desperate attempt to generate interest in these titanic struggles. But if the reader doesn't own shares in the companies in question, why should he or she care? Cable or telephone? Television or computer? Fiber optic or wireless? Most of the time, the press is satisfied with indicating winners and losers, and all too rarely willing to initiate a discussion of the underlying societal or cultural issues.

The "information highway" was first used to describe a project by the National Information Infrastructure (NII) launched by the United States government. The project made provisions for a modest public investment ($400 million) for the construction of fiber-optic networks. But it first had to create the legal and regulatory conditions needed to develop novel communications services in the fields of education and health, and a regulatory framework capable of accommodating the rapid expansion of a new interactive digital communications market. At a time when media were still clearly distinct, antitrust laws, network operating rights, and the various limitations imposed on television cable operators and phone companies all had to be updated to accommodate the new technological infrastructure, within a contemporary perspective clearly focused on the convergence of digital communication.

The expression "information highway" is unfortunate in several respects. It implies that the new system of communication has yet to be constructed, whereas it is currently in use and has been since the beginning of the eighties. Access to data banks, newsgroups, distance learning, and home shopping are in daily operation on the Internet and the French Minitel. It's true that communications channels will need to follow the increase in traffic and demand (assuming there is any) for interactive videoconferencing. But this increase in bandwidth is part of an earlier and ongoing trend. The real event, the crossing of the threshold, took place *before* the American government's project got off the ground: it began with the exponential growth in the number of Internet users that occurred between 1988 and 1991. The American government project should be considered a *response* to this social phenomenon of the expansion of "cyberculture," comparable to the one that drove the first explosion of microcomputing at the end of the seventies and start of the eighties.

Moreover, the expression connotes only the transmission speed, the physical infrastructure of communication, whereas from the social, cultural, and political point of view I want to focus on (which is of primary interest to America's citizens), the technical infrastructure is important only to the extent that it conditions the practices of communication. New modes of communication and access to information are defined by distinctness and ease of personalization, reciprocity, a nonhierarchical and hypertextual style of navigation, participation in diverse virtual communities and worlds. None of this comes through in the highway metaphor, which evokes only the transport of information or narrowly channeled mass communication, rather than interactive relationships and the creation of community.

The term "cyberspace," however, clearly indicates the opening of a communications space that is qualitatively different from those we were familiar with before the eighties. To me, it is linguistically and conceptually preferable to the terms "multimedia" or the "information highway." Indeed, my purpose in writing this book has been to develop an understanding of what cyberspace is and *could become*.

Approaching cyberspace by assimilating it into a technical infrastructure frequently masks the important fact that, in the case of digital interactive communication, functional networks are independent of physical networks. In other words, a perfectly coherent and reliable interactive communications system can function with an indeterminate number of media (radio waves, conventional telephone, cable, etc.) and encoding systems (digital, analog), with the help of the appropriate interfaces and translators. Today interactive digital communication is growing exponentially using an entire range of *existing* heterogeneous infrastructures. The continuing increase in bandwidth is only one of the keys to the growth of network traffic. Algorithms for compressing and decompressing data, which use the autonomous processing ability of intelligent network terminals (computers), represent a second, complementary method for increasing the speed of communication.

The essential point is that cyberspace, involving the interconnection of computers around the world and a communications system that is simultaneously collective and interactive, is not an infrastructure: it is a certain way of using existing infrastructures and exploiting resources and is based on an incessant distributed inventiveness that is indissolubly technical and social.

Some network operators think they have reached the apogee of enlightenment by claiming that "content is everything." But separating the "pipe" from its "content" is nothing more than a way of sharing a market (you'll sell information; we'll bill for service). The Internet—to take one well-known example—was constructed gradually and interactively, without the separation of content from the network. A large number of the files circulating on the network were software programs designed to improve the system. Moreover, beyond the pipes and the content they carry, the primary purpose of the network was, and still is, the mega-community or the innumerable microcommunities that bring it to life. The key element of cyberspace isn't the consumption of information or interactive services but participation in a social process of collective intelligence.

By assimilating cyberspace into an infrastructure, we cloak a social movement inside an industrial program. And it is indeed a social movement, for the growth of interactive digital communication wasn't decided by any multinational or government. Certainly, the American government played an important supporting role, but it was never the driving force behind the spontaneous international movement of young urban students that exploded at the end of the eighties. Aside from government funding and fee-based services offered by private companies, the growth of cyberspace rests largely on the voluntary cooperation of thousands of persons in hundreds of different institutions and dozens of countries.

The way in which cyberspace developed suggests that it is not a conventional territorial and industrial infrastructure but a self-organizing technosocial process, finalized in the short term by a categorical imperative for connectivity (interconnection being a goal in itself) that strives toward an ideal of collective intelligence, which has, to a large extent, already been implemented. The relationship between cyberspace and the city, between collective intelligence and the territory, is primarily a question of political imagination.

Articulation

Neither simple analogy, nor substitution, nor assimilation, the perspective I want to offer involves a consideration of the *articulation* of two spaces that are qualitatively very different: the territory and collective intelligence.

The territory is defined by its limits and its center. It is organized into systems of physical or geographic proximity. By contrast, every point in cyberspace is, in principle, copresent with any other, and movement can take place at the speed of light. But the difference between the two spaces is not only a function of their physical and topological properties. Contrasting social processes are also at work. Territorial institutions tend to be hierarchical and rigid, whereas the practices of cybernauts generally privilege nonhierarchical modes of relationship and fluid structures. Territorial political organizations are based on representation and delegation, whereas the technical possibilities of cyberspace make innovative forms of large-scale direct democracy practical.

To avoid any misunderstanding about "electronic democracy," I would again like to emphasize that it is not a question of enabling remote groups of people to vote instantaneously on simple proposals submitted to them by some telegenic demagogue but a question of encouraging the collective and continuous elaboration of problems and their cooperative, concrete resolution by those affected.

In articulating the two spaces, we do not eliminate territorial forms only to replace them with a style of operation that is cyber based. Rather, the goal is to compensate, at least to the extent possible, for the lassitude, inertia, and unavoidable rigidity of the territory with its real-time manifestation in cyberspace and the resolution and elaboration of urban problems through the sharing of skills, resources, and ideas.

To choose collective intelligence requires not only that we change the way cities or regions or institutions operate but also that we organize functions in cyberspace specially designed from this perspective. The following list will provide an idea of what I'm referring to:

- the dynamic representation of resources and flows
- virtual meeting places where skills, jobs, and training can be exchanged
- ecological, economic, educational, health, and other performance data, which can be consulted by anyone and updated directly with actual physical variables or activities through the use of widely dispersed sensors (which respect the anonymity of users)
- control of transport and communications systems based on real-time feedback from all users
- systems that enable users to evaluate materials and services (frequency, notifications, suggestions) together with transparent budget allocations, which implies a preference for relying on society as a whole rather than experts to measure social utility

Each of these instruments should be complemented by the use of electronic forums, where diverging viewpoints can be aired, suggestions for improvements advanced, and information and services exchanged among residents. Cyberspace projects such as this to promote collective intelligence attempt to make human groups *as conscious as possible of what they are doing together* and provide them with practical means of coordination. The goal is the ability to articulate and resolve problems using a logic of proximity and implication.

Universal access? Yes. But the expression shouldn't be understood as "access to hardware," a simple material connection—and one that will soon be inexpensive—or even "access to content" (consumption of information or knowledge distributed by specialists). Rather, we should understand it to mean universal access to the process of collective intelligence, that is, to cyberspace as an open system for the dynamic self-mapping of the real, the expression of singularities, the articulation of problems, the weaving of a social bond through reciprocal apprenticeship and the unfettered navigation of knowledge spaces. Such a perspective is an incentive not to abandon the territory so that we can lose ourselves in the "virtual" but rather to use the virtual to more fully inhabit the territory, to become complete citizens.

We "inhabit" all the environments with which we interact. We inhabit (or will inhabit) cyberspace just as we do the geographic city, and it will become a significant part of our global living environment. The development of cyberspace is a particular form of urbanism or architecture, although not physical, whose importance can only increase with time. However, the supreme architecture is really political. It involves the articulation of different spaces and their respective roles. To make collective intelligence the command post of modern society is to again choose democracy, to reactualize it by exploiting the most positive potential of our new communications systems.

PART III
PROBLEMS

CHAPTER 14

Conflict

Notwithstanding the major trends toward virtualization and universalization that were discussed earlier, the "impact" of new technologies on society and culture will be neither automatic nor predetermined. Aside from the underlying indeterminacy of sociohistorical processes, any number of contradictory interests and projects will clash with one another in cyberspace. In part 3, I want to address issues of conflict, diversity, and criticism.

Governments will clash with one another in trying to promote their industrial champions and national cultures. The conflict will be accompanied by an opposition between the interests of the state, associated with sovereignty and territoriality, and the deterritorializing and ubiquitous nature of cyberspace. Questions of censorship and cryptography—especially now that anyone can send encrypted messages across the Internet without fear they'll be deciphered—highlight the opposition between the logic of government and that of cyberculture.

We know that cyberculture constitutes an immense battlefield for communications and software manufacturers, but we must be careful about accepting at face value the war that pits powerful economic forces against one another. It conceals an underlying conflict between a purely consumer-oriented vision of cyberspace, the vision of manufacturers and merchants—in which the network is seen as a planetary supermarket and form of interactive television—and the vision of the social movement that supports cyberculture, which takes its inspiration from the exchange of knowledge, new forms of cooperation, and collective creation in virtual worlds.

The best use we can make of the tools of digital communication is, in my opinion, to effectively conjoin human intelligence and imagination. Collective intelligence is diverse; it is universally distributed, continuously enhanced, undergoes real-time synergies, and culminates in an optimal mobilization of skills. As I understand it, the goal of collective intelligence is to make the resources of large communities available to persons and small groups—not the reverse. It is a fundamentally humanist project, which encompasses, together with contemporary instruments, the great emancipatory ideals of Enlightenment philosophy. Several versions of the project for collective intelligence have been defended, however, not all of which are heading in the direction I have just outlined. Also, because its efficiency helps accelerate the ongoing mutation and thereby isolates or excludes those who are not yet a part of it, collective intelligence is ambivalent. Nevertheless, it is the only general program explicitly focused on general welfare and human development that can address the risks of a nascent cyberculture.

The Technological Future
In chapter 1, I claimed that even though society was not determined by changes in technology, cyberculture's destiny was not completely available for interpretation or incorporation by sovereign agents. It is impossible for an agent, regardless of his or her power, to control, or even understand, all the factors that contribute to the emergence of contemporary technoculture, especially when new ideas, practices, and technologies continue to issue from the most unexpected sources. Moreover, cyberculture's future can't be controlled because in the majority of instances, several agents, projects, and interpretations will be in conflict.

The acceleration of change, virtualization, and open-ended universalization are fundamental trends, most likely irreversible, which we should integrate into our reasoning and decision-making processes. Yet the way in which those trends will be embodied and absorbed in economic, political, and social life remains undetermined.

The struggle between conflicting forces and projects forestalls the illusion of the total availability of technology. In addition to the economic and material constraints that limit them, our projects must come to terms with rival projects. Yet the very fact that there is a conflict confirms the *open-ended* nature of the technological future and its social implications.

We can set up a computerized communications network in a company in such a way that a hierarchical and compartmentalized organizational structure is not only maintained but reinforced. But we can also take advantage of the occasion to promote nonhierarchical communications, enhance available skills, initiate new forms of cooperation, encourage universal access to public expression, and establish "corporate memories" that promote the accumulation and sharing of experience. Although both are technically feasible, they will be supported by different groups and give rise to power struggles and compromise.

In a school, we can restrict the communications network to implementing and promoting the use of computer-assisted learning software. We can also open up the local network to the Internet and encourage the purchase of hardware and software that will promote the autonomy and collaborative abilities of the students. Here too, contradictory educational purposes (possibly concealing infighting within the institution) can be translated into different technical configurations.

Merchants and the Future of the Absolute Market

Even within a broader perspective, the future of cyberspace involves competing projects and interests. For its inventors and early promoters, the network is a space for unrestricted interactive and community-based communication, a global instrument of collective intelligence. For others, such as Bill Gates, the chairman of Microsoft, cyberspace should become an immense, transparent global market for goods and services. This project promotes the arrival of "true liberalism" as it was imagined by the fathers of political economy, since it will exploit the technical possibility of eliminating intermediaries and making information about products and prices nearly perfect for all market agents, producers, and consumers alike.

For "content" sellers (Hollywood studios, television stations, video game distributors, data sources, etc.), cyberspace would serve as a universal data bank in which we can locate and consume, after a suitable financial transaction, any message, any piece of information, any software program, any image, any game imaginable.

For the dominant economic forces, telecommunications carriers or sellers of information, software, and services, the large questions are about the market. Is there a market for a given service? What kind of overall sales can be expected for a given category of information? When? Consultants, who generally share their clients' point of view, rarely present anything but these kinds of questions. There would be little point in emphasizing the narrow-mindedness of this point of view, even though it is obviously a legitimate one.

If Bill Gates and others interpret cyberspace as a kind of global department store, the final stage of economic liberalism, it is obviously because they are selling the tools for accessing this virtual supermarket and the corresponding instruments of transaction. The commercial interpretation of cyberspace obscures another project, however, whose objective is to redefine the market to benefit those who control certain technologies and disfavor (at least in cyberspace) traditional economic and financial intermediaries, including banks. Small producers and consumers, who may benefit from the transparency of this cybermarket, are invited to share in this goal as well as the interpretation of the phenomena it controls.

The Media Point of View

The mass media are also a powerful promoter of interpretations about cyberculture. For years television and the popular press have presented cyberspace as being infiltrated by government spy agencies and the Mafia, alarming the public with news about the ready availability of child pornography and information on Nazism and terrorism, along with fantasies of cybersex. I'd like to examine this last point for a moment. "Cybersex" is usually defined as a sexual relation that takes place remotely over a network by means of a virtual reality suit that comprises stereoscopic glasses, movement sensors, and probes on the erogenous zones. It's a form of mutual telemasturbation involving hardware that vaguely resembles an S/M outfit. Yet with the exception of a few demonstrations during specialized technology expos or artist installations— which generally involve very expensive equipment and are always public—*no one* is practicing cybersex. This doesn't prevent journalists from talking about it or well-known intellectuals from filling page after page of purple prose. Unlike cybersex, however, the Mafia, terrorists, pedophilia, and pornography in general do exist on the Net (as they do elsewhere), even though they represent only a tiny fraction of the available information. But although mobsters, terrorists, and pedophiles also use airplanes, highways, and telephones (which obviously expand their field of action), no one has thought of identifying these technological networks as tools for criminals.

The point of view propagated by the media is dictated by their interests. To get the public's attention, they need sensational news, spectacular pictures. Unfortunately cyberspace is a poor venue for such activities, since it primarily harbors distributed, asynchronous processes of collective reading and writing. There is nothing to see, nothing to show. Imagine a film sequence of someone reading, then someone writing, then another person reading—not exactly the best method for ensuring audience retention. There is no way to understand or experience what's going on in cyberspace without actively participating or by listening to the people involved in virtual communities, people who "surf the Net" and can narrate their own stories of reading and writing. Epistolary literature is not exactly a hot topic for peak viewing times. Because the reality isn't quite as sensational as the media would hope, we get terrorism, the Mafia, and cybersex.

Yet the negative connotation given to the network by some media also arises from the fact that, as I've said many times, cyberspace is an *alternative* to traditional mass media. It enables individuals and groups to find the information they want and distribute their version of the facts without going through journalists. Cyberspace encourages mutual and community-based exchange, whereas traditional media make use of one-way communication in which receivers are isolated from one another. This creates a kind of antinomy, or fundamental opposition, between the media and cyberculture, which explains the deformed reflection the media offer to the public. Obviously this doesn't prevent at least some journalists from becoming ardent users of Internet resources or the majority of the major media outlets from providing an on-line version of their services.

Government

Governments have their own point of view, one that is far-reaching and comprehensive, on the emergence of cyberspace. The narrowest approach states the problems in terms of sovereignty and territoriality. But cyberspace is fundamentally deterritorializing, whereas the modern state is based on the notion of territory. By means of the network, in-formational goods (computer programs, data, information, works of various kinds) can instantaneously move from one point on the digital planet to another without being filtered by customs offices. Financial, medical, legal, educational, and data services can also be provided to "nationals" by foreign companies or institutions (or vice versa) instan-taneously, efficiently, and nearly invisibly. The state thus loses control over an increasingly large part of cross-border economic and informa-tion flows.

Moreover, national legislation obviously applies only within a country's borders. But cyberspace can be used to easily circumvent laws regarding information and communication (censorship, copyright, ille-gal associations, etc.). All it requires is that the server providing the contested information or facilitating the prohibited communication be physically located in some "data haven" outside the country's borders for it to be beyond the reach of its jurisdiction. But now, because citizens can connect to any server in the world, providing they have access to a computer connected to a phone line, it's as if national laws regarding information and communication have become irrelevant.

Cryptography is another topic directly affecting national sovereignty. In 1991 Phil Zimmerman, an American with somewhat anarchist political beliefs, developed PGP (Pretty Good Privacy). PGP enables two correspondents on a network to securely identify each other and encrypt their messages so that they can't be broken even by the most sophisticated software running on the most powerful supercomputers. PGP is inexpensive and relatively easy to use. It incorporates the latest research in cryptology, the mathematical science that deals with encrypting and decrypting messages. After it was released on the network, the first (free) version of PGP had immediate success, and thousands of copies made their way around the world in a matter of days. PGP provides anyone with a power (complete privacy) that was formerly the exclusive privilege of the armed forces. It also insulates citizens from the kind of communications control (opening letters, phone tapping, intercepting digital messages) that police forces, even those in the most democratic countries, have practiced and continue to practice, whether for political reasons (totalitarian terrorism, surveillance, antiterrorism) or as part of the struggle against crime.

Obviously governments view the "democratization" of such powerful instruments of cryptography as an attempt on their sovereignty and security. This is why the United States government attempted to impose the use of an encryption standard to which its security forces would have a key. In the face of the public outcry, the federal government abandoned the idea of making the Clipper Chip a requirement on all American telephones and computers. Several governments, however, including France and China, require that all cryptographic technology be subject to government approval (which is very difficult to obtain). Under current legislation, the thousands of French citizens using PGP without official authorization are in possession of military weapons and could compromise national security.

In the opposing camp, the cypherpunks and crypto-anarchists (including Phil Zimmerman) are fighting to develop and maintain what they consider an important victory for citizens against the power of government. But cryptography for the masses also has many active defenders in the business community, who want to do business on the network or sell the tools for on-line transactions. It's difficult to conceive of expanding commerce in cyberspace without encrypting credit card numbers or the use of one of the competing "cybercash" systems— all of which incorporate a cryptographic module in their technical specifications. Without such measures, the risk of theft or embezzlement would certainly impede the growth of electronic commerce. It's also worth noting that banning encryption tools in a given country in no way prevents their use elsewhere by terrorists or criminals, for whom their illegality is hardly a deterrent and who can easily procure them on the network. The appearance of PGP in 1991 created an irreversible situation.[1]

Governments are obviously concerned with losing part of their sovereignty and their customary methods of surveillance. There are other, more positive points of view, however. Some experts, working for governments or international organizations, anticipate increased growth and employment following the emergence of the new economic sector of multimedia. Governments want to encourage their national industries, including software, interactive video, and on-line services, so that the positive economic and social effects of new media do not solely benefit other countries, which may hold a technological lead. Some governments perceive that the battle for economic supremacy in multimedia is complicated by the struggle for cultural influence. Consequently they attempt to reinforce their linguistic presence, enhance their informational or cultural heritage, and promote original national creativity on the network.

Although much more positive than a purely defensive attitude, the latter approach treats the concepts of culture and collective identity as givens, whereas cyberculture calls them into question. It's also possible that multimedia and online services are not simply the most chronologically recent sectors in an unchanged economy but the visible technological expression of a profound mutation of the economy itself, which requires that we also redefine concepts such as product, corporation, employment, labor, and commerce. Unlike consumer goods, the products of the new economy are personalized, interactive, actualized, coconstructed or designed by their consumers. The enterprise is virtualized, globalized, reduced to its core competencies and strategic orientations, inseparable from its network of partners (what would a "national" enterprise look like in cyberspace?). Work becomes the actualization and renewal of skills, a propensity for cooperation, rather than the execution of a prescribed task. The model of the full-time long-term employee working for a single employer is nothing but the residue of a bygone era, whereas new forms of independent labor or remuneration based on the value of skills in context have not yet become widespread. And as I noted earlier with respect to on-line commerce, the forms of competition and consumption are also changing.

Today's economic dynamism depends on the capacity of individuals, institutions, enterprises, and organizations in general to animate autonomous centers of collective intelligence (independence) and draw on the resources provided by a global collective intelligence (openness). Intelligence here should be understood as education, learning abilities (collectively and reciprocally), acquired and synergized skills, dynamic reserves of shared memory, a capacity for innovation, and receptivity to innovation. But intelligence should also be understood the way it is used in the admonition "to live in harmony," or "wisely." Collective intelligence therefore assumes the ability to create and maintain confidence, an aptitude for weaving together durable links. Cyberspace provides a powerful substrate for collective intelligence, both in its cognitive and in its social sense. As I've tried to emphasize, collective intelligence represents everything that is most positive in cyberculture on an economic, social, and cultural level. And by embracing collective intelligence, governments will be able to win back—in terms of actual power and the concerns of their populations—what they have lost through deterritorialization and virtualization.

The Public Good

Every point of view concerning the network, every interpretation of cyberculture, can be associated with a set of interests and projects. We can even say that any public expression of a point of view, any description of cyberspace, is an act that tends to accredit a specific version of the facts and promote a possible future. By helping to construct the representation of reality created by their contemporaries, experts, intellectuals, and journalists have an important responsibility. In this sense, the most important source of descriptions and interpretations of the network turns out to be the network itself. The World Wide Web is a gigantic self-referential document in which a multitude of viewpoints commingle and respond (including the fiercest criticisms of the Web). There are also countless numbers of electronic forums on the various aspects of the network, from the most ideological to the most technical. Cyberspace, which comprises those who populate it and whose voice is inevitably plural, questions itself about an identity that is currently indefinable and will be even more indefinable in the future. Because it is self-describing, the network is self-reproducing. Every map invokes a territory that is to come, and the territories of cyberspace are paved with maps delineating other maps, in endless recursion.

There is no neutral or objective approach to cyberspace, including my own. What exactly is the underlying agenda behind my own description of cyberculture? My readers are already familiar with my religion. I am profoundly convinced that *enabling human beings to conjugate their imagination and their intelligence for the development and emancipation of the individual* is the best possible use of digital technologies. This approach has a number of implications, on several levels:

- economic (the future of an economy of skills and development conceived as enhancing and optimizing human qualities)
- political (democracy that is more direct and participatory, a global, community-based approach to problems)
- cultural (collective creativity, no separation of the production, distribution, and interpretation of works of art)

In general the collective intelligence project is that developed by the first designers and defenders of cyberspace. It is the most profound aspiration of the cyberculture movement. In one sense, this project extends (and surpasses) the philosophy of the Enlightenment. It is not a question of creating a "technological utopia" but a question of strengthening the ancient ideal of the emancipation and exaltation of the human, based on today's technology. This project is reasonable for the following three important reasons.

First, collective intelligence and the technology that supports it cannot be decreed or imposed by a central power, or by disconnected administrators or experts. Its beneficiaries must also be responsible. Its operation can only be progressive, integrating, inclusive, and participatory. There is no consumer or subjugated subject in collective intelligence, or if there is, there is no collective intelligence. Consequently, since its inception, the growth of cyberspace has essentially been a spontaneous, decentralized, and participatory activity. Scientific, economic, and political power can support, promote, or at least not interfere with its development. This is what enabled the Internet, as it has for various corporate or community networks, to expand the way it has.

Second, collective intelligence is much more an open-ended problem—both in a practical and theoretical sense—than a turnkey solution. Notwithstanding the number of experiments and activities involved, it is more a question of a culture that needs to be invented than a program to be applied. Moreover, various network theorists offer sometimes divergent versions of the project, and I've already pointed out this ambivalence, both poison and cure.[2]

Third, the existence of a technical infrastructure in no way guarantees that *only* the most positive virtualities will be actualized from the point of view of human development. We can prepare the way but not determine the outcome. It will be many years before the conflict of projects and interests comes to an end. Even though cyberspace is now expanding irreversibly, the future remains open with respect to its ultimate significance for our species. Because this new communications space is sufficiently vast and tolerant, apparently exclusive projects can be realized simultaneously and may turn out to be complementary.

There are a number of obstacles that stand in the way of the project for collective intelligence. Some of them are based on misunderstanding and the overly pessimistic ideas that have been spread through frequently unfounded criticism. For this reason, in the next three chapters, I attempt to deconstruct the principal arguments supporting this critique, especially the erroneous idea that the virtual is a substitute for the real and the partial truth that cyberspace will lead only to new forms of domination.

CHAPTER 15

Critique of Substitution

Ill-founded and frequently unnecessary criticism of technology inhibits the engagement of citizens, creators, public powers, and entrepreneurs in activities that are favorable to human progress. Unfortunately such reservations leave the field open to projects that are guided solely by the search for profit and power, which are indifferent to intellectual, social, or cultural critiques. For this reason, I would like to analyze some of the arguments used in criticisms of cyberculture. I would especially like to point out the errors behind the assumption that the virtual is a substitute for the real or that telecommunications and telepresence will purely and simply replace physical movement and direct contact. The perspective of substitution sidesteps the analysis of effective social practices and appears blind to the introduction of new planes of existence, which add to previous systems and increase their complexity without replacing them.

Faced with the rapid rise of a global and destabilizing phenomenon, which challenges a number of existing assumptions, habits, and representations, I feel that as a scholar, expert, and teacher, my role is certainly not to tilt at windmills or fan the flames of anger and resentment among the public. This is apparently not the position taken by a number of intellectuals and so-called critics. Lucidity is indispensable, but it requires that we recognize the emergence of cyberculture as a phenomenon that is both irreversible and partly indeterminate. Rather than instilling fear by emphasizing aspects of cyberculture that are minimally important (cybercriminality, for example), partially relevant (cyberspace in the service of global capitalism, American hegemony, a new ruling class), or poorly understood (the virtual as a substitute for the real, the disappearance of physical space), I prefer to emphasize what is qualitatively new about the cyberculture movement and the opportunities it offers for human development. Fear is rarely a good incentive for critical thinking. Denouncing or condemning what is visibly a major component of the human future won't help us in making responsible choices.

Replacement or Complexification?

To help forestall any legitimate uncertainty, I'd like to spend some time refuting the most widespread arguments raised by some of our contemporary "critics." One of the most erroneous ideas, perhaps the most persistent, represents the substitution of the old by the new, the natural by the technological, the virtual by the real. For example, both a cultivated public and economic and political leaders fear that the rise of communications through cyberspace will replace direct human contact.

It is very rare that a new mode of communication or expression completely supplants an earlier one. Do we speak less since the invention of writing? Obviously not. However, the function of living speech has changed, since one of its purposes in purely oral cultures is now fulfilled by writing: the transmission of knowledge and tradition, the preparation of contracts, the accomplishment of important social and ritual activities. New forms of knowledge (theoretical knowledge, for example) and genres (legal codes, the novel) have come into being. Writing hasn't caused speech to disappear; it has complexified and reorganized the system of communication and social memory.

Has photography replaced painting? No. Painters continue to paint. Attendance at museums, exhibitions, and galleries is high, and collectors continue to purchase artwork to hang on their walls. On the other hand, it's true that painters, graphic artists, engravers, and sculptors are no longer the sole manufacturers of images, as they were in the nineteenth century. The ecology of the icon has mutated; painters have had to reinvent painting—from impressionism to neo-expressionism, abstraction to conceptual art—so that it would regain its uniqueness and become irreplaceable in the new environment created by modern industrial methods for producing and reproducing images.

Has cinema replaced theater? Certainly not. Cinema is an autonomous genre, with its own materials, history, rules, and codes. And authors, actors, theaters, and audiences continue to populate the contemporary cultural landscape. The rise of television certainly affected cinema, but it didn't destroy it. We frequently watch films on television, and TV networks coproduce films.

Telecommunications and Transportation

Did the development of telephony lead to fewer face-to-face contacts and a drop in transportation? No, quite the contrary. The development of the telephone and the automobile occurred in *parallel* and to their mutual benefit. The more telephones there were, the greater the amount of urban traffic. A relation of substitution does exist between them, since if your telephone network were to fail, the likelihood and severity of traffic jams would probably increase as well. However, the most significant historical trend is toward the simultaneous increase of the instruments of telecommunications and transportation. At a finer level of detail, sociological studies have shown that the people who make and receive the greatest number of phone calls are also those who travel the most and have the greatest number of direct contacts with others. These studies confirm our intuitive apprehension of the world that surrounds us. For an older person, alone and with limited mobility, the telephone, sitting comfortably on its lace doily, rarely rings. The active businessman or -woman, however, running from one meeting to the next, can often be seen with a cell phone glued to his or her ear, in the back of a taxi, in an airport lounge, or on a street corner. The telephone's success is an eloquent demonstration that telecommunications and physical movement go hand in hand.

Cyberspace has not yet provided the kind of statistics that are available for traditional telecommunications. What do the numbers say?

Users of cyberspace are primarily young, well educated, and urban; they are students, teachers, and researchers, often working in the sciences, high-tech industries, business, and contemporary art. Yet these individuals are among the most mobile and most sociable elements of the population. The typical Internet user is running from one international conference to another and is frequently a regular member of one or more professional communities. My personal experience and that of the cybernauts I know personally confirms this assumption. Those whose electronic mailbox is overflowing and who actively surf the Web are the same people who travel frequently and have a fairly large circle of acquaintances. We prepare our conferences, meetings, or "physical" exhibitions using the tools supplied by the Internet. We can even extend a seminar in the form of an electronic conference. Interestingly, it is the students with access to the Internet who are most involved in extra-curricular activities and travel.

I'm the first to admit that certain Net junkies spend their nights in front of a computer, playing networked video games, participating in on-line discussions, or surfing interminably from Web page to Web page. These exceptions confirm the rule of nonsubstitution. The image of the terminal-man for whom space has been abolished, immobile, glued to the screen, is nothing more than a fantasy dictated by fear and misunderstanding of phenomena that are subject to deterritorialization, globalization, and a *general increase in relationships and contacts of all kinds.*

The assumption of simple substitution contradicts all the available empirical and statistical evidence.[1] It is disturbing to discover that the five most recent books by a critical thinker such as Paul Virilio are based on a fantasy, which is shown to be false by the simple observation of the world around us. Nicholas Negroponte, no prophet of misfortune but an enthusiastic specialist of high-tech marketing, wrote of "the transition from atoms to bits"[2] in *Being Digital*, the substitution of information for matter, of the virtual for the real. It's worth pointing out that in the economic sphere, the amount of international trade, measured in tonnage (thus, in atoms), has continued to increase over the last fifteen years, in spite of the telecommunications revolution and growth of cyberspace.

Had Negroponte claimed that the control of information and strategic skills, or the ability to process and distribute digital data efficiently, now *determined* the production and distribution of goods, I would have agreed with him. But the *transition* from atoms to bits is an outrageous simplification that borders on the absurd.

The Universe of Choice and Rise of the Virtual

Museums are often evoked as a paradigm in the cultural sphere, and I'd like to take a look at what's been said about them. We fear (or wish) that "virtual museums" will replace "real" museums, that the use of on-line museum services or Web sites devoted to the arts might interfere with the streams of visitors to the buildings that house original artworks. Although disturbed by the possible disembodiment of art or its relation to the world in general, I want to emphasize again that the pale digital copy available on the Internet will never replace the sensual richness of the physical artwork. Anyone can confirm this for him- or herself. But if we look at history, we find that the multiplication of printed reproductions, art magazines and monographs, museum catalogs, films and television programs on ceramics, painting, and sculpture, haven't reduced, but have on the contrary increased, museum attendance.

The historical trend can be summarized as follows: as information accumulates, circulates, and proliferates, it is used more efficiently (rise of the virtual), and the variety of physical objects and locations with which we are in contact (rise of the actual) increases. Yet our informational universe expands more rapidly than our universe of concrete interactions. In other words, the rise of the virtual leads to a rise in the actual, but the first develops more quickly than the second. This is the reason for the impression that we are being deluged with data, messages, and images, for the perception of a lag between the virtual and the real. For the majority of us, our experience of art is based more on reproductions than on original works of art. However, the popularity of large exhibitions, the increase in the number of museums, and the ease with which we can travel have given us access to more original works of art than were available to the average European or American in the nineteenth century. Museum holdings available on the Internet or CDs will increase our opportunities to discover and understand a vast range of artworks, which will encourage viewers to examine the materiality of painting and sculpture directly. Unfortunately current debates about substitution obscure the genuine aesthetic and cultural opportunities currently available to us, which involve new modes of creation and reception and emerging artistic genres based on the tools of cyberspace.

There is a generalized fear that although the tools for communication and virtual interaction "save time," they will produce generations of loafers, whose relation to understanding is negligent and hurried, and who are apt to forget the effort involved in discovering the "real" world and its sensible wealth. Once again, moralizing platitudes have replaced observation and thought. For example, consider the effective consequences of some of our "timesaving" technological innovations. Refrigerators, freezers, dishwashers, vacuum cleaners, household robots, detergents, and other products have not resulted in "women's liberation." Statistics show that the average time spent by women cleaning house and cooking was nearly the *same* before and after the introduction of household appliances. Domestic technologies haven't saved us any time; they have improved our standards of living, hygiene, and cleanliness. They have also made it easier for women to work outside the home, which has resulted in *increasing* the overall amount of time they work rather than reducing it.

Recent studies on urban and land development have shown that increases in the speed of circulation almost never lead to a reduction in the time spent traveling between home and work. On the contrary, the average travel time (between twenty minutes and a half hour) has remained remarkably constant. However, people have rearranged the physical location of their home to extend their universe of choice. It still takes a half hour to get to work, but we now also have ready access to entertainment, schools, restaurants, and meeting places.

I'd like to risk generalizing from this last example. We assume that speed and virtualization (technological in origin) will save us time. In reality they enable us to spend the same amount of time, and sometimes even more, exploring and exploiting a much greater informational, relational, or concrete space. Speed (and the virtual is fundamentally a mode of speed) does not cause space to disappear; it results in the metamorphosis of the unstable and complicated system of human spaces. Each new vehicle, each new aspect of acceleration, invents a topology and quality of space that are added to their predecessors, articulates and reorganizes the global economy of space. There is a highway space, a topology of railroads, a global aviation and airport network, a specific cartography of telephony. Each of these spatiotemporal strata engenders its own system of proximity, zones of density, black holes, uncharted regions. But highways, planes, telephones, and the Internet haven't caused the disappearance of local roads or trails (fully accessible to the casual stroller or traveler); they have transformed their function.

New Planes of Existence

The root for the idea of substitution in interpreting technological change seems to be the difficulty in grasping, imagining, or conceptualizing the appearance of new cultural forms, whose dimensions are far beyond what the human world is accustomed to. During biological evolution, when the eye first appeared, it took over some of the functions of touch and smell, but it primarily resulted in the appearance—by developing the vague sensitivity to light that existed on certain parts of the animal's skin—of a previously nonexistent universe of form and color, *the experience of vision.*

As with the appearance of new organs, the major technological inventions not only enable us to do "the same things" more quickly, better, or on a greater scale but also allow us to do, feel, or organize ourselves differently. They lead to the development of new functions while requiring that we readjust the overall system incorporating the previous functions. The problematic of substitution prevents us from conceiving, accepting, or promoting that which is qualitatively new, that is, new planes of existence that are virtually supported by technical innovation.

Let's return to the example of photography. The fact is that photography has not replaced painting, even though it has made the "optical capture" of a scene easier and faster than when using a brush or colored pigments, even though it has democratized the ability to fix an image, including for those of us who have no ability to draw. There is indeed a purely substitutional aspect to photography: more people are making similar images, which are easily reproducible, and doing so faster than before and without any special skills. Yet photography has also brought about the deployment of new functions for the image. It has done so not only within the space it created for itself—through the phenomena of echoing and differentiation—but within painting as well, which it was thought photography would replace. With the invention of photography, the visual arts have undergone a split (photography on one side, painting on the other) that implies other forms of interaction (cinema, animation, interactive images, virtual reality) and the introduction of a new graphic perspective. Every bifurcation in the technology of the image results in the development of potentialities already latent in the older visual arts. We reinterpret older painting based on our experience of photography, cinema, and contemporary painting. We reinterpret cinema based on virtual reality. There is a continuous growth and enrichment that accompanies the process of emergence and radical exposure.

Concerning Loss

Technological innovation engenders growth, the actualization of latent virtualities. It also contributes to the creation of new planes of existence. It adds complexity to the foliation of aesthetic, practical, and social spaces. This does not mean things will never disappear. There is no longer a blacksmith in every village or horse manure on the streets of our cities. Something has been lost. The habits, skills, and modes of subjectivation that were adapted to the old world are no longer adequate. Technological change results in suffering almost by necessity. To spurn such change, deny it, fail to recognize it, or recognize only its negative aspects, will simply add to the inevitable pain. So how can we limit our suffering? By remaining lucid in the face of such transformation, by participating in its movement, by committing ourselves to a learning process and seizing the opportunities for human growth and development.

The development of cyberspace is not going to miraculously "change our life" or resolve contemporary economic and social problems. It will, however, expose new planes of existence in our

- modes of relation: interactive and community-based any-to-any communication within collectively and continuously reconstructed information spaces
- modes of understanding, apprenticeship, and thought: simulation, nonhierarchical navigation within unrestricted information spaces, collective intelligence
- literary and artistic genres: hyperdocuments, interactive works, virtual environments, distributed collective creation

The communications systems, modes of knowledge, and genres characteristic of cyberculture will not simply replace earlier modes and genres. They will influence them and force them to find their specific "niche" in the new cognitive ecology.[3] The overall result will be (is already) a complexification and reorganization of the economy of information, knowledge, and art.

There is little doubt that cyberculture will become the center of gravity of the twenty-first century's cultural galaxy. But the claim that the virtual will become a substitute for the real, or that we will no longer be able to distinguish between them, is simply a bad joke, which fails to grasp any of the significance of the concept of virtuality. If the virtual encompasses digital information and communication, the proposition is absurd: We will continue to eat, make love, travel, produce and consume material goods, and so forth. If the virtual is understood in its philosophical sense, it must be paired with the actual or actualization, in which case it then becomes a particularly fecund mode of reality. What about the anthropologically virtual? Has language, the power of lies and truth, the first virtual reality to take us beyond the here and now, sheltered from immediate sensation, caused us to lose touch with reality, or has it exposed us to new planes of existence?

CHAPTER 16

Critique of Domination

It is claimed that the network has a tendency to reinforce existing centers of scientific, military, and financial power and that e-commerce will experience dizzying growth within the next few years. Nevertheless it would be wrong, as some critics have done, to reduce the coming of the new communications space to the acceleration of economic globalization, the intensification of traditional domination, or even the appearance of unknown forms of power and exploitation. Cyberspace can also be used for personal or regional development, for participation in the emancipating and unrestricted processes of collective intelligence. Moreover, these two perspectives are not necessarily mutually exclusive. Within an increasingly interconnected and interdependent universe, such opposing tendencies can even sustain each other. In this chapter, I'll show how the dynamism of cyberspace is closely associated with the initiation and continuation of a genuine dialectic between utopia and business.

Powerless Media

The coming virtual apocalypse is currently fashionable. Some professional critics attempt to portray the forces at work as if they were puppets in a traditional morality play: Ladies and Gentlemen. Here we have the poor, the hungry, the downtrodden, the people of the Third

World (feeling guilty yet?). Here we have their nefarious antagonists: technology, capital, finance, multinational corporations, government. It is true that governments can have a marginal effect on the conditions—whether favorable or unfavorable—affecting the growth of cyberspace. However, they have been quite obviously powerless to guide the development of a communications system that is now inextricably linked with the functioning of global economy and technology.

As I noted in chapter 1, the majority of the great technological innovations of the last few years have not been determined by the large corporations that generally serve as the targets of whiny critics. In fact, inventiveness has been unpredictable and widely distributed. The best example is the success of the World Wide Web. When at the beginning of the nineties, the press and television spoke of the multimedia industry and the information highway, they introduced the key players: the U.S. government, the CEOs of large computer software and hardware companies, cable and telecommunications carriers. However, a few years later, we are forced to acknowledge that although the big "media" players did manage a number of mergers and acquisitions, they did not significantly affect the development of cyberspace. Between 1990 and 1997, the principal revolution in planetary digital communication came from a small group of researchers at CERN, in Geneva, who developed the World Wide Web. It was the social movement of cyberculture that made the Web the success it is today by propagating a means of communication and representation that corresponded to its methods and ideals. Critics watch television, with its talking heads, whereas the important events take place within widely distributed and invisible processes of collective intelligence, which necessarily escape the attention of conventional media. The World Wide Web wasn't invented or distributed or driven by the big media players such as Microsoft, IBM, AT&T, or the American army, but by cybernauts themselves. As such we are justified in questioning the progressive character of contemporary criticism, which presents us with the same old demoralizing bugbear and ignores or denigrates a social movement.

The New Virtual Class

They promised you utopia, electronic democracy, shared knowledge, and collective intelligence. You ended up being dominated by a new virtual class,[1] *made up of media (film, television, video games), software, electronics, and telecommunications magnates, flanked by the designers, scientists, and engineers who oversee the construction of cyberspace. Ultraliberal and anarchist ideologues, high priests of the virtual, serve as their spokesmen and interpreters, justify their power.*

Another, Third World or European, version of this paranoid narrative presents the development of cyberspace as an extension of the American military, economic, and cultural empire. By now any attempt to highlight the positive aspects of cyberculture or the opportunities it offers for human development are suspected of serving the interests of an emerging virtual capitalism or new systems of global domination, whether technofascist, cyberfinancial, or American-liberal.

Obviously there is some truth in this type of analysis, but only some. The enormous mutation of contemporary civilization will bring with it a redefinition of the nature of military, economic, political, and cultural power.[2] Some forces will gain power, others will lose power, while newcomers are beginning to stake out positions that didn't even exist before the emergence of cyberspace. On the virtual checkerboard, the rules haven't been completely determined yet. Those who manage to define them to their own advantage will win and win big. At present, and in spite of the considerable instability of the situation, the dominant centers of military and financial power are well placed to increase their influence. Still, we must also be alert to the openness and indeterminateness of the process of technosocial change currently under way.

In spite of our legitimate suspicions, one fact remains: a group, or an individual, regardless of geographic or social origin, even though they are nearly without economic means, can, providing they have access to a minimum of technical skills, turn cyberspace to their own account and acquire data, make contact with others, participate in virtual communities, or distribute information they feel worthy of interest to a broad public. These new forms of communication will continue—and may even grow richer—as long as cyberspace grows. There is little risk in predicting that they will continue to develop in the future.

This simple fact demolishes the gloomy analyses developed by Arthur Kroker and others, who take science fiction to be gospel truth and offer their public nothing but clumsy imitations of Baudrillard. Unfortunately, Baudrillard himself imitates situationist radicalism, without the cold, clear, objective intelligence of Guy Debord or the quasi-mystical force of Vaneigen. The situationists denounced the *spectacle*, the type of interpersonal relation crystallized by the media: centralized nodes distribute messages to isolated receivers who are unable to respond. Within the spectacle, the only form of participation is imaginary. However, cyberspace offers a style of communication that is inherently free of media intervention, since it is communal, nonhierarchical, and reciprocal.

It is television, not the virtual, that systematizes our inability to act and the feeling of unreality that follows. Television forces me to share the same eye, the same ear as millions of other people. The shared perception is generally a good indicator of reality. However, although they organize this shared perception, the media do not enable those who perceive the same "reality" to communicate with one another. Although we're all equipped with the same type of eardrum, we're unable to hear one another. We watch the same performance but don't recognize ourselves. Moreover, the televisual retina and tympanum are cut off from our brain, isolated from the sensorimotor feedback loop. We see through someone else's eyes, without the ability to turn our gaze where we wish. Perhaps it is only through telepresence systems, which enable us to effectively control the extension of our organs at a distance, that we can bring this negative judgment to bear on television. Television is an important source of reality because it organizes a shared perception. But it is also a powerful source of unreality because perception is disconnected from systems of action, and participation in the sensorimotor loop is one of the surest signs of the real. This is the reason for the feeling of bewilderment it produces. With television we participate as a group, but without the ability to contribute, in someone else's dream or nightmare.

In cyberspace, distribution from some central point is replaced by interaction within a situation, a universe of information, which everyone can explore as he or she chooses, can modify or stabilize (reestablishing the sensorimotor loop). Cyberspace harbors negotiations about meaning, processes of mutual recognition among individuals and groups *through* the activity of communication (cooperation and debate among the participants). These processes do not exclude conflict. They obviously imply the presence of persons or groups who are not always driven by the best intentions, for cyberspace provides all the diversity, complexity, and, yes, even the cruelty of the real, a thousand miles from the pat, predetermined scenarios that issue from the media. It is much harder to succumb to manipulation within a space where everyone can transmit information and where contradictory information can collide than in a system where the transmitting centers are controlled by a minority.

It's true that virtual reality promises a better illusion than that provided by film or television. However, there is no evidence that anyone has ever confused an interactive virtual world with "true" reality. No matter how much we enjoy virtual worlds or want to return to them, it is impossible to forget their fictional character, which constantly comes back to haunt us (weight of the headset, poor quality of the image, reaction time needed to calculate images in real time, etc.). We can't say the same for televised news or armed forces videos, which are presented as images of reality and often perceived as such.

With respect to a situationist morality, the true danger haunting cyberspace is the large-scale transposition to the "information highway" of a media-based mode of communication. Cyberspace doesn't seem to be heading down that path, but the option remains open, and powerful forces are pushing it in that direction. The worst scenario would be for the Internet to be replaced by a gigantic system of "interactive television."

Unfortunately a number of so-called radical critical intellectuals, dazed by their long hours of interaction with the media, confuse the virtual with their personal feeling of unreality and are unable to distinguish among different types of communication. Subject to a range of conservative influences, they exacerbate the public's vague fears, preventing it from grasping the strategic alternatives available.

The Dialectic of Utopia and Business

Have network utopians become the pawn of economic or military powers? Yes and no. It's true that the Internet came into being as a result of a decision by the American army. The system was initially designed to enable labs across the country to access supercomputers at a handful of locations. The project was immediately altered, however, and from its origin, the Internet served primarily as a means for researchers to correspond with one another. Out of the powerful machine provided by the military authorities, the first designers and users created a nonhierarchical communications space.

We also know that the dispersed structure of the network was designed to be optimally resistant to enemy nuclear attacks. Yet this dispersed structure currently serves a cooperative and decentralized purpose. Thus, paradoxically, the Internet may be "anarchist" not *in spite* of its military origin but *because* of it.

After its initial military phase, the network grew through the efforts of researchers and students who were engaged in "utopian" practices of communal exchange and took a democratic approach to their relation to knowledge. Its growth wasn't determined by any large corporation or government, even though the American government and some large companies were part of the movement. The public wasn't really aware of this cooperative and spontaneous construction of a gigantic international electronic mail system until the end of the eighties. From then on, it has become increasingly the focus of commercial interests, which have fought to sell access, organize its structure, pillage its content, and transform it into a new space for advertising and merchandising. Cyberspace is big business.

What began as a utopian movement characterized by its generosity was turned into a mother lode for business. But conversely, the economic forces at play in cyberspace and the commingling of e-commerce with other activities of production and exchange are now such that the existence and continued development of the Internet (with the characteristics typical of its communications system) are almost certainly guaranteed. Business has solidified and made irreversible what an actual utopia had begun to construct. I should add that many of the innovative entrepreneurs active in the cybereconomy are also visionaries. At the large global technology conferences and virtual markets, the long hair of the cyberpunks and the eccentric outfits of hip designers commingle with the close-cropped hair and three-piece suits of businessmen. Sometimes its hard to tell who is who.

Business is not inherently evil. The growth of cyberculture has been supported by a dialectic between utopia and business, in which each plays against the other. Until now there have been no losers. Cultural and social projects cannot be radically separated from the economic constraints and dynamism that make them possible.

The cyberculture movement is one of the engines of contemporary society. Governments and the multimedia industries follow it as best they can, slamming on the brakes in their attempt to slow down what they perceive to be the "anarchy" of the Net. Following the well-worn dialectic between utopia and business, merchants exploit the fields of existence (and thus consumption) opened by the social movement and acquire new sales arguments from network activists. Similarly, the social movement benefits from its "co-optation" because business solidifies, standardizes, makes credible, and institutes ideas and ways of doing that in the not so distant past were viewed as science fiction or fantasy.

After the fall of totalitarianism in the East, some Central European intellectuals said, "We fought one another for democracy . . . and wound up with capitalism." Cyberculture activists could make this phrase their own. Fortunately, capitalism is not completely incompatible with democracy, nor collective intelligence with the planetary supermarket. We are not required to choose between them: the dialectic of utopia and business, the interplay of industry and desire.

CHAPTER 17

Critique of Criticism

Functions of Critical Thought

Cyberculture has been supported by an expansive social movement that both reveals and engenders a profound change in civilization. Critical thought's role is to comment on cyberculture's orientation and modalities of implementation. In particular, progressive criticism can help determine the most positive and original aspects of the ongoing evolution. In this way, it will help ensure that we do not turn the mountain of cyberspace into a molehill: the perpetuation of the media on a larger scale or the fulfillment of the on-line planetary supermarket.

However, a great deal of so-called critical discourse is blind and conservative. Because it misunderstands the ongoing transformations, it fails to produce original concepts adapted to the specificity of cyberculture. It criticizes "the ideology (or utopia) of communication" without distinguishing between television and the Internet. It creates fear of a dehumanizing technology, ignoring the fact that what is key is how we choose among technologies and the different uses we make of them. It deplores the growing confusion between the real and the virtual without realizing that virtualization, far from being a derealization of the world, is rather an extension of human abilities.[1] The absence of a vision of the future, the abandonment of imagination and anticipation, have discouraged others from getting involved and left the field open to commercial propaganda. It is essential, including for criticism itself, that we begin to critique a "critical genre" that has been destabilized by the new ecology of communication. It is important that we begin to question mental habits and reflexes that are increasingly unable to address the key issues facing contemporary society.

Totalitarianism and Detotalization

The idea that the development of cyberspace is a threat to civilization and humanist values is largely based on the confusion between universality and totality. We are suspicious of anything that presents itself as being universal because universalism has, almost always, been advanced by conquering empires and absolute rulers, whether their domination was temporal or spiritual. Yet cyberspace, at least until now, has been more receptive than dominating. It's not an instrument of hub-and-spoke distribution (like the press, radio, and television) but an interactive communications system among human collectives and a tool for interconnecting heterogeneous communities. Those who see in cyberspace the danger of "totalitarianism" have simply diagnosed the problem incorrectly.

It's true that governments are responsible for reading our correspondence, stealing data, manipulating information, or spreading misinformation in cyberspace. None of this is radically new. The same thing happened in the past and continues in the present using other means: the post office, the telephone, traditional media, even physical theft. Digital communication tools are more powerful, however, and can be used to do harm on a much larger scale. But we must also recognize that powerful encryption tools, which are now accessible to the individual user,[2] provide a partial solution to these threats. I again want to emphasize that television and the press are far more effective instruments of manipulation and misinformation than the Internet because they can impose "a" vision of reality to which there is no response, criticism, or confrontation among divergent positions. This was very much the case during the Gulf War. In contrast, a diversity of sources and unrestricted discussion are inherent in cyberspace, which is essentially "uncontrollable."

By associating cyberculture with a "totalitarian" threat, we misrepresent its nature and the processes governing its growth. It is true that cyberspace constructs a universal space, but as I've tried to show, it is a universal without totality. This is really the crux of the problem, for isn't it the current process of detotalization that has professional critics terrified? Doesn't their condemnation of the new means of interactive, nonhierarchical communication echo the old yearning for law and order? Haven't we demonized the virtual to preserve unchanged a reality that has been instituted and legitimized by the "common sense" of government and media?

Those whose job is to manage limits and borders are threatened by unrestricted, nonhierarchical, multipolar communication. The guardians of good taste and quality and their intermediaries and spokesmen realize that their positions are threatened by the establishment of increasingly direct relationships between producers and consumers of information.

Texts circulate throughout the world by means of cyberspace without ever having passed through the hands of a publisher or editor. Soon the same will be true of music, film, hyperdocuments, interactive games, and virtual worlds. Since it is now possible to publicize new ideas and experiments without going through the peer review process of a specialized journal, the entire system of control in science has been called into question. The appropriation of knowledge will continue to free itself of the constraints imposed by educational institutions, since the living sources of knowledge will be directly accessible and individuals will be able to become a part of virtual communities devoted to cooperative learning. Doctors will now compete with medical databases, newsgroups, and virtual mutual assistance groups formed by patients afflicted with the same disease. Those who hold contemporary positions of power and possess needed "skills" now find themselves threatened. But if the individuals and groups who serve as intermediaries can successfully reinvent themselves and become leaders in the process of collective intelligence, their role may become even more important in the new civilization than it was in the past. However, if they insist on clinging to their old identity, in all likelihood their situation will deteriorate.

Cyberspace will not change power relations and economic inequalities. But to take a simple example, power and wealth are not distributed or used in the same way in a caste society, with its hereditary privileges and restrictive corporate monopolies, and a society in which all citizens are equal before the law, and in which free enterprise and the struggle against monopolies are encouraged. By increasing market transparency and promoting direct transactions between buyers and sellers, there is little doubt that cyberspace will facilitate a "liberal" evolution of the economy of information and knowledge, and very likely in the overall operation of the economy.[3]

Should this liberalism be understood in the noblest sense of the word: the absence of arbitrary legal constraints, opportunities given to talented individuals, free competition among a large number of small producers within a fully transparent market? Or will it serve as a mask, an ideological pretext for domination by large communication conglomerates, which will make life as difficult as possible for small producers and stifle diversity? These two paths are not mutually exclusive. And the future will probably incorporate elements of both, whose ultimate proportions will depend on the strength and orientation of the social movement of cyberspace.

Contemporary criticism believes that it is based on a denunciation of impending "totalitarianism" and serves as spokesman for the "disenfranchised," whose opinion it rarely heeds. In fact, the critical pseudoelite is nostalgic for a totality that it was able to control. But this sentiment has been negated, reversed, and projected onto a terrifying other: cybercultural humanity. Concern over the decline of semantic closure and the dissolution of controllable totalities (experienced as a breakdown of culture) mask a defense of power. None of this will delay the invention of the new civilization of the universal-through-contact, nor will it be of any use in giving it the most human orientation possible. We should try to grasp cyberculture from within, through the multiform social movement that drives it, and in terms of the originality of its communications systems, by identifying the new forms of the social bond it weaves within the heavily populated silence of cyberspace, far from the monotonous clamor of the media.

From Progressive to Conservative

Skepticism and a systematic critical attitude played a progressive role in the eighteenth century, a period of political absolutism during which freedom of expression had yet to be won. Today skepticism and criticism may have changed sides. Such attitudes are increasingly becoming an excuse for blasé conservatism, if not more reactionary positions. In their pursuit of the spectacular and the sensational, contemporary media continue to present the most depressing aspects of current events, constantly put politicians on the spot, and denounce the "dangers," the negative effects, of economic globalization and technological development. They play on people's fear, one of the easiest sentiments to provoke. Today's thinkers no longer need to spread panic by aligning themselves with the attitudes of the popular press and television but to once more analyze the world from a new perspective, offer a deeper form of understanding, new mental horizons for their contemporaries, who are bathed in the media's discourse. Should intellectuals and professional thinkers therefore abandon all critical perspective? Certainly not. But it is important to bear in mind that the critical attitude in itself, whether simple nostalgia or parody of the wide-ranging criticism of the eighteenth and nineteenth centuries, is no longer a guarantee of cognitive openness or human progress. We must carefully distinguish between a mechanical criticism, which is media based, conventional, conservative, an excuse for existing forms of power and intellectual laziness, and *actualized* criticism, which is imaginative, forward-looking, and supportive of the social movement. Not all criticism is thoughtful.

The Ambivalence of Power

The contemporary acceleration of the race toward the virtual and the universal can't be reduced to the "social impact of new technologies" or the arrival of a specific form of domination, whether economic, political, or social. We sense that there is something narrow, limited, even absurd in such proposals. The race is a movement of civilization as a whole, a kind of anthropological mutation in which—along with the growth of cyberspace—coexist demographic growth, urbanization, the densification of transport networks (and the correlative increase in traffic), technoscientific development, the (unequal) elevation in the level of education, the omnipresence of the media, the globalization of production and exchange, international financial integration, the rise of large transnational political groups, and the evolution of ideas tending toward a growing awareness of humanity and the planet.

At the same time, the self-alienation and power of our species is increasing. By complexifying and intensifying its relations, by discovering new forms of language and communication, multiplying its technical means, we become *even more human*. This gradual invention of the essence of humanity, which is still taking place, does not ensure a uniformly glorious future or an increase in happiness. Universalizing and virtualizing tendencies are accompanied by increased inequality between rich and poor, between central regions and disenfranchised zones, between participants in the universal and the dispossessed. They interrupt or marginalize secular transmission, weaken local cultures, which are part of the most precious heritage of our species, and violently destabilize the imagination that organizes subjectivity. They encourage *re*territorialization, a return to particular interests, a fixation with identity. In one sense, the *global civil war* that runs from ghetto rebellion to fundamentalist insurrection expresses the heartache of a humanity that is unable to unite without sacrificing itself.

The passage to the virtual is a detour, an accumulation that portends more numerous and more powerful actualizations. The fear of a "derealization of the world" is unfounded.

The open, untotalizing universal of cyberculture welcomes and enhances singularities, offers access to expression to the greatest number. Because they are outmoded, the fear of control, totalitarianism, and uniformity does a poor job of identifying a target it would be best to look for among traditional media and authoritarian and hierarchical social forms.

But cyberculture's positive potential, although it may lead to new human abilities, cannot ensure peace or happiness. In becoming more human, we would be wise to remain more vigilant. For humans alone are cruel and their cruelty is proportional to their humanity.

CHAPTER 18

Answers to Common Questions

To conclude, I would like to discuss some of the principal problems concerning the development of cyberculture, without claiming to have a solution to those problems. If sociohistorical processes are indeed fundamentally open, indeterminate, if they are continuously reestablished and reinvented, no verbal solution, no theoretical response can ever circumscribe them. Answers, which are always provisional, belong to the sociotechnological process as a whole, that is, to each one of us, depending on the scale and orientation of our possibilities for action, without implying that any of us is capable of global or final control. I have selected four questions for which there are no "answers." They address the content and significance of cyberculture in its entirety. The first question, "Does cyberculture produce exclusion?" is obviously of prime importance in a global society in which exclusion (the contemporary form of oppression, social injustice, and poverty) is one of the principal ills. The second, "Is the diversity of languages and cultures threatened?" is directed toward the absence of totalization typical of cyberculture, which has been discussed here. The third, "Is cyberculture a synonym for chaos and confusion?" assumes the absence of totalization but questions its possible negative content. The fourth question, "Does cyberculture break with the values of modern Europe?" enables me to demonstrate again that cyberculture extends and realizes the ideals of the Enlightenment and the great European current of thought concerning human emancipation. I will suggest, however, that while continuing these traditions, cyberculture implies a radical renewal of political and social thought and provokes a metamorphosis of the very concept of culture.

Does Cyberculture Produce Exclusion?

We often feel that the development of cyberculture could be an additional factor in creating inequality and exclusion among classes within a society as well as between rich and poor nations. This risk is real. Access to cyberspace requires costly communications and computational (computers) infrastructures for developing regions. The acquisition of the skills needed for putting together and maintaining server clusters represents a nonnegligible investment. Assume for a moment that the access points to the network and the equipment needed for viewing, producing, and storing digital information are accessible. The "human" obstacles, however, still remain to be overcome. These are the institutional, political, and cultural constraints on community-based, nonhierarchical, interactive forms of communication. Finally, people need to overcome their feelings of incompetence and disqualification in the face of new technologies.

Three different answers can be given to the question of exclusion. They are not conclusive solutions, of course, but they do enable us to relativize the problem and put it into perspective.

1. *We need to look at trends rather than figures about connectivity.*

In 1996 there were 1,500 people connected to the Internet in Vietnam. This sounds like a very small number given the overall population of the country. But there will be ten times as many by 2000. In general the rate of growth of connections to cyberspace reflects a rate of social appropriation greater than all previous communications systems. The post had existed for centuries before most people were able to send and receive letters on a regular basis. The telephone, invented at the end of the nineteenth century, currently reaches approximately 30 percent of the world's population.

The number of people participating in cyberculture has increased exponentially since the end of the eighties, especially among the young. In some parts of America and in the Nordic countries, more than 50 percent of the population is connected to the Internet. Whole regions and countries are planning their entry into cyberculture, and especially the most dynamic among them (Asia and the Pacific Rim, for example). The number of those excluded from cyberspace will continue to decline.

2. *It will become increasingly easy and less expensive to get connected.*

Although commonplace, feelings of incompetence are becoming increasingly less justified. The establishment and maintenance of telecommunications infrastructures by individuals and organizations require fairly high-level skills. However, once we know how to read and write, the *use* of cyberspace by individuals and organizations requires little in the way of technical understanding. The mechanisms for accessing and navigating are becoming increasingly user-friendly, especially since the development of the World Wide Web at the beginning of the nineties.

Moreover, the hardware and software needed to connect are becoming cheaper. To help reduce subscription and telecommunications costs, governments can try to encourage competition among access providers and telecommunications carriers. The key point in all this is the cost of the local call. In North America it is part of the standard monthly rate. We pay the same if we are on-line for five minutes or five hours. European rates for local calls, however, are based on the amount of time spent on-line, which discourages use of the Internet, bulletin boards, or any other form of interactive networked communication.[1]

3. *Any advance in communications systems necessarily results in some form of exclusion.*

Every new communications system results in exclusion. There were no illiterates before the invention of writing. Printing and television introduced the division between those who publish or appear in the media and everyone else. As I have noted, it is estimated that only 30 percent of the world population has a telephone. None of these facts constitutes a serious argument against writing, printing, television, or the telephone. We do not condemn writing or telecommunications because some people can't read or don't have a telephone. On the contrary, it encourages us to develop primary education and expand our telephone networks. The same should be true of cyberspace.

In a more general sense, every universal produces exclusion. The universal, even though it is totalizing in its classic forms, never incorporates the whole. A universal religion has its believers and its heretics. Science tends to disqualify other forms of knowledge or whatever it considers irrational. "The rights of man" result in infractions and regions of lawlessness. Older forms of the universal excluded by separating those who participate in the truth, meaning, or a given form of empire from those who are cast into the shadows: barbarians, infidels, the ignorant, and so forth. The universal-without-totality cannot escape the rule of exclusion. However, here it is no longer a matter of adherence to meaning but of connectivity. The excluded individual is disconnected. He or she doesn't participate in the relational and cognitive density of virtual communities and collective intelligence.

Cyberculture is a wholesale collection of heresies. It mixes citizens and barbarians, the (so-called) ignorant and the learned. Unlike the separation that typified classic forms of the universal, its borders are fluid, moving, and provisional. Still, the disqualification of the excluded is no less appalling.

It is important to bear in mind that because of the way they were constructed, the older universals inevitably resulted in exclusion. A universal religion or science must assume that earlier or parallel systems are mistaken. However, the movement of the universal-through-contact is inclusive: it asymptotically approaches generalized interconnectivity.

What, then, can we do? Clearly, we need to reduce the difficulty and cost of getting connected by whatever means appropriate. But the problem of "universal access" can't be reduced to the technological and financial dimensions we generally hear about it. We can't overcome our feeling of inferiority by sitting in front of a computer with a user-friendly interface. We need to be prepared to participate actively in the process of collective intelligence, which is what makes cyberspace so valuable. Our new tools should serve to enhance our culture, skills, resources, and local projects; they should help people participate in mutual assistance collectives, cooperative learning groups, and so on. In other words, from the perspective of cyberculture, as well as of more traditional approaches, affirmative policies to combat inequality and exclusion must strive for *increased autonomy* for those involved. However, such policies must also avoid creating new forms of dependency caused by the consumption of information or communication services designed and produced for purely commercial or imperial ends, which all too frequently disqualify the traditional forms of knowledge and skill belonging to disadvantaged social groups and regions.

Language and Culture in Cyberspace

English is currently the de facto language standard on the network. What's more, American institutions and companies are the largest producers of information on the Internet. The fear of cultural domination by the United States is not without some basis in fact. However, the threat of uniformity is not as serious as it may appear at first sight. The technological and economic structure of communication in cyberspace is very different from that of film or television. In particular, the production and distribution of information is much more accessible to individuals and groups with fairly modest means. The issue of cultural diversity makes sense only in the context of the specific structure of the communications tools used by cyberculture.

One of the most important aspects of the emergence of cyberspace is the development of an alternative to the mass media. By mass media I am referring to communications systems that distribute organized, programmatic information from a central point to a large number of anonymous, passive, and isolated receivers. The traditional press, film, radio, and television are typical examples of such media. Cyberspace, however, is based not on such a hub-and-spoke model of distribution but on one of shared spaces where everyone can have his say and use what interests her. It is the kind of information market where people can interact and where the requester takes the initiative. The most likely "centers" in cyberspace are the data servers that power Web sites. But a server is better suited to a store, a place where one responds to demand by offering a variety of goods, than to a one-way distribution system.

Of course it would be technically and politically possible to implement a mass-media model in cyberspace. But I feel it is more important to recognize the new potential made available through generalized interconnectivity and the digitization of information. I've summarized this potential in the following four sections.

The End of the Monopolies of Public Expression

Everyone, regardless of who they are, can now have access to the technical means to reach an enormous international public at little cost. Anyone (group or individual) can distribute fiction, journalism, or personal accounts of current events in a given field.

The Growing Variety of Modes of Expression

The modes of expression available for communicating in cyberspace are already highly varied and will be increasingly so in the future. From simple hypertext to multimodal hyperdocuments and digital video, to interactive simulations and performances in virtual worlds, new ways of writing images, new rhetorics of interactivity will be invented.

The Increasing Availability of Instruments for Filtering and Navigating the Information Deluge

Automated or semiautomated instruments for filtering, navigating, and orienting oneself within the network will enable us to quickly obtain the information we feel is most pertinent to our needs. This does not necessarily imply the introduction of electronic blinders, since the "most pertinent" can be, if I so desire, that which is far removed from my customary interests. This new capacity for detailed filtering and automated retrieval within large masses of information will, in all likelihood, mean that "abstracts" targeted at the least common denominator within an anonymous mass will become increasingly less useful. They will shift the "center of informational gravity" toward the individual or group in search of information.

The Development of Virtual Communities and
Remote Interpersonal Contacts Based on Affinity

The persons who populate and enrich cyberspace are its principal source of wealth. Access to information is probably less important than the ability to communicate with experts, agents, the direct witnesses of subjects that interest us. Yet cyberspace makes it increasingly easy to meet individuals on the basis of their skills and fields of interest. Involvement in open communities focused on research, practice, and debate will be the best antidote against dogmatism and the unilateral manipulation of information. Moreover, cyberspace promotes our integration in "virtual communities" independently of physical and geographic barriers.

Cultural diversity in cyberspace will be directly proportional to the active involvement and quality of the contributions of representatives of different cultures. It's true that certain material infrastructures (telecommunications networks, computers) and a minimum of skill are required. Nonetheless the important point is that the political, economic, and technological constraints to the global expression of cultural diversity have never been as weak as they are in cyberspace. This does not mean that such barriers are nonexistent, simply that they are much lower than with other communications systems.

Even a cursory investigation of the World Wide Web reveals an uncontrollable abundance of information and forms of expression from all over the world (even though much of it comes from North America) and the most varied intellectual outlooks. Not only do our complaints about the lack of diversity fail to correspond to reality, but more importantly, *there is no one to complain to.* Cyberspace contains whatever people put into it. Maintaining cultural diversity depends primarily on the initiative of each of us, and possibly the support that governments, foundations, international organizations, and NGOs are capable of providing to artistic and cultural projects.

What about language? That English is the lingua franca of the Net (as it is in the scientific community, business, tourism, etc.) is unquestionably a handicap for those who are not native speakers. It should be pointed out, however, that the very existence of a lingua franca is an advantage for international communication. It would appear difficult to do without one. But why English? Aside from American dominance in economic, military, and cultural fields, we must acknowledge that English (spoken in England, the United States, Canada, Australia, South Africa, and elsewhere) is currently the predominant language on the Internet. In order of demographic importance, however, English ranks third, following Chinese and Hindi, but the rate of connectivity in China and India is still quite low. (Incidentally, English is followed by Spanish, Russian, and Arabic.)

But the fact that English is dominant on the network does not mean that it is the only language used. At present we can find information on the Internet in hundreds of different languages. Large numbers of texts are available in French, Spanish, Portuguese, German, Italian, and other languages. Virtual communities have been created on the basis of linguistic affinity, which intersect and complicate existing topical affinities.

Restrictions to the maintenance and growth of linguistic diversity are essentially technical. With current standards, accented characters using the Roman alphabet (such as French or Spanish) are somewhat disadvantaged compared to those that are unaccented (such as English). Non-Roman alphabets (Cyrillic, Greek, Arabic, Hebrew, Korean) are at an even greater disadvantage. Languages based on ideographic characters, such as Chinese and Japanese, are the most disadvantaged of all. Although the limitations are real, these languages can certainly be graphically represented. Moreover, with the advances in research (especially research involving the use of nonalphabetic writing systems) and the next round of standards, we will be able to send text in Russian or Chinese as easily and transparently as we now do English.

Aside from these minor technical difficulties, there are no obstacles to linguistic diversity on the Internet—other than the lack of initiative or activity among speakers of a given minority language. It seems to me that if we are going to retain a healthy dose of humility and mutual respect, the proper attitude is to treat all languages as minority languages, especially one's own. English itself is a minority language when compared to Chinese, or within the community of French speakers. Although French was once an imperial language, native speakers must accustom themselves to thinking of French as a minority language. Regional variants, dialects, patois, and oppressed or disappearing idioms are also minority languages that need to be defended and protected— both on and off the Net. Bear in mind that the vitality of expression in cyberspace is not "Anglo-Saxon" but American. The Québécois, for example, are French-speaking Canadians. Yet "although French-speaking Canadians represent only five percent of the global Francophone population, thirty percent of all pages published in French on the Internet are from Québec."[2]

What, then, should we do? Common sense suggests that authors shouldn't publish exclusively in English if they're not native speakers. They should provide the original version of their text, possibly with translations into languages other than English. Similarly, if we are trying to reach an international audience, it is advisable to provide an English version along with the original text, to ensure the broadest possible distribution.

Chaos and Confusion

Can we have confidence in the information found in cyberspace when anyone can feed the network without an intermediary or fear of censorship, when no government, institution, or moral authority guarantees the quality of the available data? Because no official selection or hierarchy is there to help us find our way through the informational deluge of cyberspace, are we simply witnessing a form of cultural dissolution rather than progress? And will such dissolution ultimately serve only those who are already able to find their bearings in cyberspace, that is, those who are privileged by their education, background, or personal intellectual networks? Although these questions are legitimate, they are based on false premises.

It is certainly true that no centralized authority guarantees the value of the information available throughout the network as a whole. Nonetheless Web sites are produced and maintained by persons or institutions who sign their contributions and defend their validity before the Internet community. To give one obvious example, the contents of a university site are guaranteed by the university that hosts it. As in the print world, on-line reviews and newspapers are the responsibility of an editorial committee. Companies warrant the information they provide, and maintaining their reputation on the Web is as important as with other forms of communication. Government information is obviously controlled by the government.

Virtual communities, mailing lists, and newsgroups are often moderated by responsible individuals who filter messages based on their quality or relevance. It is also not uncommon for system administrators to be employed by public organizations (universities, museums, ministries) or institutions interested in maintaining their reputation (large corporations, associations, etc.). These system administrators, who wield considerable "regional" power in cyberspace, can block information (newsgroups, spam, etc.) that contradicts the ethics of the network (netiquette) from the servers they administer: calumny, racism, direct encouragement to violence, sexual solicitation, systematic dissemination of irrelevant information. This also explains why there is so *little* of this type of information or behavior on the network.

There is also a kind of public opinion that operates on the Internet. The best sites are often cited or discussed in reviews, catalogs, and indexes (on-line or print). Large numbers of hypertext links point to such "good" services or sites. There are proportionally far fewer links to sites with little or declining informational value.

The operation of the network essentially depends on the suppliers and users of information in a public space. It rejects hierarchical (and therefore opaque), global, and a priori control—which could be a possible definition of censorship or the totalitarian control of information and communication. We cannot at the same time expect freedom of expression and the a priori selection of information by an organization that supposedly knows what is true and good for everyone, whether that organization is journalistic, scientific, political, or religious.

But what of the chaos and confusion, the diluvian character of information and communication in cyberspace? Aren't those who are without strong personal or social reference systems disadvantaged by this? This fear is only partly legitimate. The profusion of information, the lack of an overall a priori order, do not prevent persons or collectives from getting their bearings and constructing their own hierarchies, creating their own structure. What has disappeared for good are selections, hierarchies, and structures of knowledge that claim to be valid for everyone at all times, namely, the totalizing universal. As I mentioned earlier, on-line indexes and increasingly sophisticated search tools provide Web users with a degree of local and provisional order within the global disorder. Moreover, it is wrong to conceive of cyberspace as being populated by isolated individuals lost within masses of information. The network is primarily an instrument of communication among individuals, a virtual place where communities help their members learn what they want to know. Data are only the raw material of a living and highly developed intellectual and social process. Finally, all the collective intelligence in the world isn't going to eliminate the need for personal intelligence, individual effort, and the time required to learn, search, evaluate, and become a part of a community, even a virtual one. The network will never do our thinking for us—for which we should be thankful.

Cyberculture and Fundamental Values

In contrast to postmodernism's assumption of a decline of the ideas of the Enlightenment, I believe that cyberculture can be considered a legitimate (although distant) heir of the progressive project of eighteenth-century philosophy. Cyberculture confirms our participation in communities where debate and argument can take place openly. In keeping with the morality of equality, cyberculture encourages a fundamental reciprocity in human relations. Throughout its development, it has seriously encouraged the exchange of information and knowledge, which Enlightenment philosophers considered as the principal engine of progress.

Therefore, assuming we were ever modern,[3] cyberculture wouldn't be postmodern but firmly situated in the tradition of revolutionary and republican ideals of liberty, equality, and fraternity. However, in cyberculture, these "values" are embodied in concrete technological systems. In the age of electronic media, *equality* is realized as the ability of each of us to transmit to everyone else; *liberty* is objectivized in encryption software and cross-border access to a multiplicity of virtual communities; *fraternity* takes the form of global interconnectivity.

Thus, far from being resolutely postmodern, cyberspace appears as a kind of technical materialization of modern ideals. In particular, the contemporary evolution of information technology constitutes an astonishing realization of the Marxist objective of appropriation of the means of production by the producers. Today "production" consists essentially of simulating and processing information, creating and distributing messages, acquiring and transmitting knowledge, and integrating it in real time. Personal computers and digital networks place the tools of economic activity in the hands of individuals. What's more, if as the situationists claim, the spectacle (the media system) is the pinnacle of capitalist donination,[4] then cyberspace represents a genuine revolution, since it enables—or will soon enable—anyone to bypass publishers, producers, distributors, and intermediaries in general and disseminate text, music, virtual worlds, or any other product of their intellect. In contrast to the television viewer's isolation and in ability to respond, cyberspace provides the conditions for direct, interactive, collective communication.

The quasi-technical realization of the ideals of modernity immediately reveals their partial and incomplete (rather than mediocre) character. It is obvious that neither personal computing nor cyberspace, no matter how widespread their use becomes, will ever resolve by themselves the principal problems of life in society. Yes, they will help us realize new forms of universality, fraternity, community, and the reappropriation of the instruments of production and communication by the people. However, at the same time, they rapidly, and often violently, destabilize economies and societies. While destroying the old, they participate in creating new powers, which are less visible and more unstable, but not less virulent.

Cyberculture appears to be the partial solution to the problems of the previous epoch, but it is itself an immense field of problems and conflicts for which no sense of global resolution can clearly be identified. Among the social forms that have been most severely implicated, our relation to knowledge, work, money, democracy, and the state needs to be reinvented.

In one sense, cyberculture continues the great tradition of European culture; in another, it transmutes the very concept of culture. In the conclusion, I would like to turn my attention to this subject.

Cyberculture, or
The Simultaneous Tradition

Far from being a subculture of Net fanatics, cyberculture expresses an important change in the very essence of culture. In keeping with the arguments I've developed in this book, the key to the culture of the future is the concept of the universal-without-totality. Here the "universal" signifies the virtual presence of humanity to itself. The universal shelters the here and now of the species, its point of encounter, a paradoxical here and now without any clearly assignable place or time. For example, a universal religion is supposedly addressed to all humanity and reunites it virtually through its revelation, its eschatology, its values. Likewise, science is supposed to express (and apply to) the intellectual progress of everyone without exception. Thinkers are the delegates of the species, and the triumphs of the exact sciences are those of humanity as a whole. Similarly, the goal of a cyberspace that we claim to be universalizing is to interconnect all speaking bipeds and encourage their participation in the collective intelligence of the species within a ubiquitous environment. In a completely different way, science and universal religions open virtual spaces where humanity can encounter itself. Although it fulfills an analogous function, the way cyberspace reunites people is much less "virtual" than that of science or the great religions. Scientific activity implicates everyone and is addressed to everyone through the intermediary of a transcendental subject of knowledge in which every member of the species participates. However, by ensuring that humanity is present to itself, cyberspace implements a technology that is real, immanent, and graspable.

And what about *totality*? Using my terminology, totality is the *stabilized unity of meaning associated with diversity*. That this identity might be organic, dialectic, or complex rather than simple or mechanical changes nothing: it is always totality, an encapsulating semantic closure. However, in bringing about humanity's virtual presence to itself, *cyberculture does not impose a unity of meaning*. That is the principal argument I have been trying to defend in this book.

The categories discussed are typified by three major historcal stages:

- small, self-centered oral societies that lived a nonuniversal totality
- "civilized" imperial societies that used writing and led to the development of a totalizing universal
- cyberculture, which corresponds to the concrete globalization of societies and invents a universal without totality

We need to bear in mind that the second and third stages do not cause earlier stages to disappear; they relativize them through the addition of other dimensions.

Initially, humanity was composed of a multitude of dynamic cultural totalities or traditions, mentally self-centered, which obviously didn't prevent them from interacting with others or being influenced by them. Humans are tribal members par excellence. There are few expressions of archaic culture that are open to all human beings without exception. Neither their laws (no "rights of man"), their gods (no universal religions), their knowledge (no experimental procedures or reasoning that can be reproduced anywhere), nor their technology (no networks or global standards) is universal.

As discussed earlier there were few authors in these cultures. But the closure of meaning was ensured through transcendence, by example and the decisions of ancestors, by a tradition. True, there was no way of recording this information. But cyclical transmission from generation to generation ensured its duration over time. The capacities of human memory limited the extent of the cultural treasury to the memories and knowledge of a group of elders. They were living totalities, but not universal.

Following this, in "civilized" societies, the conditions of communication brought about by writing led to the practical discovery of universality. Writing, then printing, brought with them the possibility of the indefinite extension of the social memory. This openness to the universal took place simultaneously in time and space. The totalizing universal reflects the inflation of signs and fixation of meaning, the conquest of territories and the subjection of people. The first universal was imperial, state controlled. It was imposed on top of cultural diversity. It tended to dig out a layer of being that was everywhere always identical, supposedly independent of us (the universal constructed by science) or associated with a given abstract definition (the rights of man). Yes, our species now exists in and of itself. It comes together within strange virtual spaces: revelation, the end of time, reason, science, law ... From the state to written religion, from religion to the networks of technoscience, universality is affirmed and takes shape, but almost always through totalization, the extension and maintenance of unique meaning.

Cyberculture, however, the third stage in this evolution, maintains universality while dissolving totality. It corresponds to the moment when our species, through economic globalization and densification of the communications and transportation networks, tends to form only a single global community, even if that community is characterized by inequality and conflict. Alone in its genus within the animal kingdom, humanity reunites its entire species in a single society. But at the same time, and paradoxically, the unity of meaning bursts apart, perhaps because it is beginning to be realized in practice, through effective contact and interaction. Connected to the universe, virtual communities constantly construct and dissolve their dynamic, emerging, immersed, drifting micrototalities among the turbulent currents of the new deluge.

Traditions were deployed in the diachrony of history. Interpreters—temporal operators, smugglers across evolutionary lines, bridges between future and past—reactualized memory, simultaneously transmitted and invented ideas and forms. The great intellectual and religious traditions patiently built hypertext libraries to which each new generation added its nodes and links. Sedimented collective intelligences, the church and the university stitched the centuries together. The Talmud was a profusion of commentaries on commentaries, where yesterday's wise men conversed with those who preceded them.

Far from uprooting the pattern of "tradition," cyberculture turns it at a forty-five-degree angle, in perfect synchrony with cyberspace. Cyberculture embodies the horizontal, simultaneous, purely spatial form of transmission. Its ability to provide temporal connections is an afterthought. Its principal activity is connecting in space, constructing and extending rhizomes of meaning.

Before you lies cyberspace with its teeming communities and the interlaced ramification of its creations, as if all of humankind's memory were deployed in the moment: an immense act of synchronous collective intelligence, converging on the present, a silent bolt of lightning, diverging, an exploding crown of neurons.

1. The Impact of Technology

1. See Mark Johnson and George Lakoff, *Metaphors We Live By* (Chicago: University of Chicago Press, 1980).

2. My comments caricature, but only slightly, the thesis developed by Gilbert Hottois in *Le Signe et la Technique* (Paris: Aubier-Montaigne, 1984).

3. How do institutional structures and material technologies convey ideas? This is one of the principal research topics of the "mediologic" project, begun by Régis Debray. See, for example, his *Cours de médiologie générale* (Paris: Gallimard, 1991), *Transmettre* (Paris: Odile Jacob, 1997), and the review *Les Cahiers de médiologie*.

4. I develop this point at greater length in my *Les Technologies de l'intelligence* (Paris: Seuil, 1993). See also the research done on the new anthropology of science and technology, for example, Bruno Latour, *Science in Action* (Cambridge: Harvard University Press, 1985).

5. See Heidegger's well-known article "The Meaning of Technology," in *The Question concerning Technology, and Other Essays*, ed. William Lovitt (New York: Harper and Row, 1977), which has engendered a number of intellectual followers among philosophers and sociologists of technology and critics of contemporary life in general.

6. Is technology always on the side of "instrumental reason"?

7. The parallel between electronics and nuclear science is developed by Derrick De Kerckove in *The Skin of Culture* (Toronto: Sommerville Press, 1995).

8. A bulletin board system (BBS) is a community-based communications system that uses computers connected over ordinary phone lines. A virtual

community is a group of persons who correspond with one another by means of interconnected computers. A hypertext is a reconfigurable and fluid digital text. It is composed of basic building blocks connected by links that can be explored in real time on-screen. The notion of a hyperdocument generalizes to all categories of signs (stationary images, animations, sound, etc.) the principle of the mobile networked message, which characterizes hypertext. The World Wide Web gathers into a single immense hypertext or hyperdocument (comprising images and sound) all the documents and hyperdocuments that feed into it. For a detailed description of the significance of cryptography, see the section on government attitudes toward cryptography in chapter 16, especially concerning the various conflicts of interest and interpretation.

9. Pierre Lévy, *Collective Intelligence*, trans. Robert Bononno (New York: Plenum Press, 1997).

10. A good description of the feedback process can be found in Joël de Rosnay, *L'Homme symbiotique* (Paris: Seuil, 1995).

2. The Technical Infrastructure of the Virtual

1. Processing power is generally measured in millions of instructions per second (MIPS).

2. Storage capacities are measured in bits (an elementary encoding unit that is either 1 or 0) or bytes (8 bits). The byte corresponds to the amount of space needed to encode an alphanumeric character. A kilobyte (KB) is 1,000 bytes. A megabyte (MB) is 1,000,000 bytes. A gigabyte (GB) is 1,000,000,000 bytes.

3. Source: IBM.

4. Jaron Lanier is one of the best-known representatives of this type of research.

5. Bill Buxton is the best-known representative in the field.

6. HiTime, or Hypermedia Time-Based Structuring Language.

7. VRML, or Virtual Reality Modeling Language. The current VRML standard used on the WWW governs the exploration of three-dimensional models with a mouse rather than through immersion, which involves the use of stereoscopic glasses and data gloves.

3. Digital Technology and the Virtualization of Information

1. The philosophical and anthropological aspects of the virtual are discussed in my *Becoming Virtual: Reality in the Digital Age*, trans. Robert Bononno (New York: Plenum Press, 1998).

2. There are other meanings of the term in optics, mechanics, and other disciplines. In addition to my *Becoming Virtual*, the concept is analyzed in René Berger's *L'Origine du future* (Paris: Le Rocher, 1996), especially the chapter entitled "Le virtuel jubilatoire," and in Jean-Clet Martin's *L'Image virtuelle* (Paris: Kimé, 1996).

3. For example, an image can be broken down into pixels. Each pixel in a color image is represented in the computer by five numbers: two numbers for the coordinates of the dot and three numbers for the intensity of each of three elementary components of color. This type of encoding can result in information loss, however. The greater the "resolution" of the encoding, the fewer the losses. An image can be encoded in 256 pixels (256×5 numbers) or 1,024 pixels ($1,024 \times 5$ numbers). There will be less information loss in the 1,024-pixel image. Beyond a certain level of resolution, however, the loss of information is no longer perceptible to the human eye.

4. Note that the photograph, as a material object on paper, does not actually contain the image. To the ant walking across the photograph's surface or the mouse nibbling away at a corner of the paper, it's not a blossoming cherry tree. Strictly speaking, the photograph is a physical medium for various pigments, whose arrangement is interpreted as a blossoming cherry tree by our mind or, to put it differently, by the "calculations" performed by our central nervous system.

Actuality of the Virtual

1. That the virtual image represents the sensible embodiment of the world of ideas postulated by philosophy is one of the arguments of Jean-Clet Martin's interesting book *L'Image virtuelle*.

5. Cyberspace

1. My definition of cyberspace is close to, although more restrictive than, that provided by Esther Dyson, George Gilder, Jay Keyworth, and Alvin Toffler in their "Magna Carta for the Knowledge Age," *New Perspective Quarterly*, (fall 1994): 26–37. For these authors, cyberspace is "the land of knowledge," the "new frontier," whose exploration "can be civilization's truest highest calling."

6. The Universal without Totality

1. For a discussion of large technological systems, see Alain Gras and Sophie Poirot-Delpech, *Grandeur et Dépendance: Sociologie des macro-systèmes techniques* (Paris: PUF, 1993).

7. The Social Movement of Cyberculture

1. See Didier Gazagnadou, *La poste à relais: La diffusion d'une technique de pouvoir à travers l'Eurasia* (Paris: Kimé, 1994).

2. A bulletin board system, or BBS, is a group communications system that uses computers connected to one another through modems and the public telephone network.

3. Christian Huitema, *Et Dieu créa l'Internet* (Paris: Eyrolles, 1996).

4. Based on a remark by Paul Soriano on the BBS of *l'Atelier*, one of the most important virtual communities in France.

5. Kevin Kelly, *Out of Control* (New York: Addison-Wesley, 1994); Joël de Rosnay, *L'Homme symbiotique.* The concept of collective intelligence is discussed at length in my *Collective Intelligence* and *Becoming Virtual: Reality in the Digital Age*, trans. Robert Bononno (New York: Plenum Trade, 1998).

6. An intranet is a network that uses standard communications protocols (TCP/IP) and the services and software typical of the Internet (Web browsers, e-mail, newsgroups, file transfer) within an organization or network of organizations.

8. The Sound of Cyberculture

1. A number of demonstrations, exhibitions, and conferences have been devoted to the arts of the virtual. Two of the most important international events are Ars Electronica, which is held in Linz, Austria, and the International Symposium of Electronic Arts (ISEA), organized by the International Society for Electronic Arts, which is held in a different city every year (Helsinki in 1994, Montreal in 1995, Rotterdam in 1996, Chicago in 1997, Paris in 2000, etc.). Exhibition catalogs and conference proceedings provide a good introduction to these works, their authors, and the theories of the arts of the virtual.

9. The Art of Cyberculture

1. Through the use of HTML, the current World Wide Web (WWW) standard.

2. This too will take place on the WWW through the use of tools such as VRML and Java.

3. Umberto Eco, *The Open Work* (Cambridge: Harvard University Press, 1989).

4. See Gilles Deleuze and Félix Guattari, *A Thousand Plateaus: Capitalism and Schizophrenia*, trans. Brian Massumi (Minneapolis: University of

Minnesota Press, 1987), and *What Is Philosophy? (European Perspectives)* (New York: Columbia University Press, 1996).

11. Education and the Economy of Knowledge

1. Gordon Davies and David Tinsley, *Open and Distance Learning: Critical Success Factors*, proceedings of the International Conference, Geneva, 10–12 October 1994.

2. Specialists in educational policy recognize the essential role of high-quality and universally accessible elementary education in determining the general educational level of a population. Elementary education affects all children, whereas secondary, and especially higher, education affects only some of them. However, secondary and higher public education, which are far more expensive than elementary education, are financed by everyone's tax dollars. The discrepancy is particularly alarming in poor countries. See Sylvain Lourié, *École et tiers monde* (Paris: Flammarion, 1991).

12. The Knowledge Tree

1. The knowledge tree, or skill tree, is a registered trademark of Trivium S.A. The trees operate with Gingo software, also developed by Trivium. See Michel Authier and Pierre Lévy, *Les Arbres de connaissances*, with a preface by Michel Serres (Paris: La Découverte, 1992). A new paperback edition with an afterword was issued by La Découverte in 1996.

13. Cyberspace, the City, and Electronic Democracy

1. See Pierre Veltz's remarkable book on economic geography entitled *Mondialisation: Villes et territoires* (Paris: PUF, 1996).

14. Conflict

1. See Jean Guisnel, *Cyberwars: Espionage on the Internet*, trans. Gui Masai, with a foreword by Winn Schwartau (New York: Plenum Press, 1997).

2. For example, in *Out of Control*, Kevin Kelly outlines an approach to collective intelligence based on biological models. Joël De Rosnay, in *L'Homme symbiotique*, introduces the perspective of a symbiotic being assembled by the Network (the "cybionte"). Joseph Rheingold, in *Virtual Communities* (New York: Addison-Wesley, 1993), provides a more political and community-based approach to collective intelligence. And in my own *Collective Intelligence*, I've tried to show that true collective intelligence enhances singularity. Collective intelligence has

become the focus of a humanist project for civilization, one that ultimately serves the person.

15. Critique of Substitution

1. See Marie-Hélène Massot, *Transport et télécommunications* (Paris: INRETS-Paradigme, 1995). Massot's book provides a complete international bibliographic analysis of the relationship between telecommunications and mobility.

2. A bit is the basic unit of information in the mathematical theory of communication.

3. On the concept of cognitive ecology, see my *Les Technologies de l'intelligence*.

16. Critique of Domination

1. On the rise of the virtual class, see Arthur Kroker and Michael A. Weinstein, *Data Trash: The Theory of the Virtual Class* (Montreal: New World Perspectives, 1994).

2. A very good description of this redistribution of power is provided in the work of Alvin Toffler. See, for example, *Powershift: Knowledge, Wealth, and Violence at the End of the Twenty-First Century* (New York: Bantam Books, 1990), and *War and Anti-war: Survival at the Dawn of the Twenty-First Century* (New York: Little Brown, 1993).

17. Critique of Criticism

1. See my *Becoming Virtual*, where this question is treated from the point of view of anthropology.

2. See the section on government in chapter 14, "Conflict." See also Jean Guisnel, *Cyberwars: Espionage on the Internet*.

3. According to Bill Gates, the most important aspect of cyberspace will be its ability to serve as the "ultimate market." See *The Road Ahead* (New York: Penguin Books, 1995).

18. Answers to Common Questions

1. Internet connections are generally charged at a flat monthly rate, regardless of the number of hours spent on-line. But the *local* connection between the user's wall outlet and the Internet service provider (ISP) is billed by the conventional telecommunications carrier.

2. Brunot Oudet, "Le multilinguisme sur Internet," *Pour la science*, no. 235 (May 1997): 55.

3. See Bruno Latour, *We Have Never Been Modern*, trans. Catharine Porter (Cambridge: Harvard University Press, 1993).

4. See Guy Debord, *The Society of the Spectacle*, trans. Donald Nicholson-Smith (New York: Zone Books, 1995).

Index

Pierre Lévy is a philosopher who has devoted his professional life to understanding the cultural and cognitive impacts of digital technologies and to promoting their best social uses. He is the author of several books, including *World Philosophy: Market, Cyberspace, Consciousness; Collective Intelligence;* and *Becoming Virtual: Reality in the Digital Age.* He is a professor of cyberculture and social communication at the University of Quebec.

Robert Bononno has been a French translator for fourteen years and has published book-length translations and numerous shorter works of fiction and nonfiction. He is adjunct professor in the Translation Studies program of New York University, where he has taught courses on technical translation and computer technology for translators, and at the Graduate School of the City University of New York, where he has taught courses on terminology and translation.